ITCA Information Technology Certified Associate

COMPUTING
FUNDAMENTALS
Study Guide

Part of the ITCA Stackable Certificate Program
www.isaca.org/itca

ISACA®

About ISACA

For more than 50 years, ISACA® (www.isaca.org) has advanced the best talent, expertise and learning in technology. ISACA equips individuals with knowledge, credentials, education and community to progress their careers and transform their organizations, and enables enterprises to train and build quality teams. ISACA is a global professional association and learning organization that leverages the expertise of its 145,000 members who work in information security, governance, assurance, risk and privacy to drive innovation through technology. It has a presence in 188 countries, including more than 220 chapters worldwide.

Disclaimer

ISACA has designed and created *Computing Fundamentals Study Guide* (the "Work") primarily as an educational resource for professionals. ISACA makes no claim that use of any of the Work will assure a successful outcome. The Work should not be considered inclusive of all proper information, procedures and tests or exclusive of other information, procedures and tests that are reasonably directed to obtaining the same results. In determining the propriety of any specific information, procedure or test, professionals should apply their own professional judgment to the specific circumstances presented by the particular systems or information technology environment.

Reservation of Rights

ISACA

1700 E. Golf Road, Suite 400
Schaumburg, IL 60173, USA
Phone: +1.847.660.5505
Fax: +1.847.253.1755
Contact us: https://support.isaca.org
Website: www.isaca.org

Provide Feedback: To comment, please visit https://support.isaca.org.
Participate in the ISACA Online Forums: https://engage.isaca.org/onlineforums

Twitter: http://twitter.com/ISACANews
LinkedIn: www.linkedin.com/company/isaca
Facebook: www.facebook.com/ISACAGlobal
Instagram: www.instagram.com/isacanews/

ISBN 978-1-60420-656-2
Computing Fundamentals Study Guide
Printed in the United States of America

Acknowledgments

The *Computing Fundamentals Study Guide* reflects the collective effort of many volunteers. Special thanks go to:

Lead Developer

Patrick Conlan, CCNP, CCSP, CCDP, CCNA, CCNA Wireless, CCDA, MCSA, MCSE, MCSE-Security, CEH, CHFI, Security+, A+, IP Assurance, USA

Expert Reviewers

Hafiz Sheikh Adnan Ahmed, CGEIT, Certified COBIT 5 Assessor, Certified DPO, UAE
Jessica Denny, USA
Daniel Johns, CRISC, CMgr, FIML,Australia
Jason Otani, CISM, USA

Board of Directors

Tracey Dedrick, Chair, Former Chief Risk Officer, Hudson City Bancorp, USA
Rolf von Roessing, Vice-Chair, CISA, CISM, CGEIT, CDPSE, CISSP, FBCI, Partner, FORFA Consulting AG, Switzerland
Gabriela Hernandez-Cardoso, Independent Board Member, Mexico
Pam Nigro, CISA, CRISC, CGEIT, CRMA, Vice President–Information Technology, Security Officer, Home Access Health, USA
Maureen O'Connell, Board Chair, Acacia Research (NASDAQ), Former Chief Financial Officer and Chief Administration Officer, Scholastic, Inc., USA
David Samuelson, Chief Executive Officer, ISACA, USA
Gerrard Schmid, President and Chief Executive Officer, Diebold Nixdorf, USA
Gregory Touhill, CISM, CISSP, President, AppGate Federal Group, USA
Asaf Weisberg, CISA, CRISC, CISM, CGEIT, Chief Executive Officer, introSight Ltd., Israel
Anna Yip, Chief Executive Officer, SmarTone Telecommunications Limited, Hong Kong
Brennan P. Baybeck, CISA, CRISC, CISM, CISSP, ISACA Board Chair, 2019-2020, Vice President and Chief Information Security Officer for Customer Services, Oracle Corporation, USA
Rob Clyde, CISM, ISACA Board Chair, 2018-2019, Independent Director, Titus, and Executive Chair, White Cloud Security, USA
Chris K. Dimitriadis, Ph.D., CISA, CRISC, CISM, ISACA Board Chair, 2015-2017, Group Chief Executive Officer, INTRALOT, Greece

Page intentionally left blank

Table of Contents

LIST OF FIGURES

Chapter 1. Computing Basics

Page intentionally left blank

Introduction

The *Computing Fundamentals Study Guide* is a reference guide designed to assist individuals in acquiring a base-level knowledge of information technology (IT) in the areas of basic computing, networks, virtualization and security. It helps individuals prepare for the ISACA® Computing Fundamentals certificate exam, one of the components of the ISACA Information Technology Certified Associate™ certification program.

Topics in this guide are presented at a foundational level and include basic terminology, concepts, examples, general practices and explanations. The *Computing Fundamentals Study Guide* provides learners knowledge in the following main content areas:

- Basic data concepts
- Hardware components of computing devices
- Software components of computing devices
- Data and file systems
- Networking concepts
- Virtualization concepts
- Security concepts

Every chapter contains the following:

- **Learning objectives**—Knowledge that learners should understand after reading the chapter
- **Sections**—Knowledge content
- **Knowledge checks**—Quiz on the chapter knowledge

Page intentionally left blank

COMPUTING
FUNDAMENTALS
An ISACA Certificate

Chapter 1:

Computing Basics

Computing Basics

Computing Basics

1.1 Learning Objectives

After completing this chapter, learners will be able to:

1. Define basic data concepts.
2. Identify hardware components and technology.
3. Identify common computing hardware devices.
4. Distinguish between major operating systems.
5. Identify common file types by their extension.
6. Describe differences among types of memory.
7. Perform mathematical conversions.
8. Describe the data life cycle.

1.2 Introduction

It is hard to imagine life without computers. Computers can increase accuracy and efficiency as well as perform complex tasks faster than humans. For many, the advent of personal computers eliminated geographic barriers and provided access to a wealth of information. Enterprises design and build networks to connect computers together and share local network resources like printers and storage devices. Servers on enterprise networks share even more information and provide services for authentication, addressing and name resolution. The largest-scale network for sharing information is the Internet. The Internet allows you to post videos, pictures and talk to friends.

Computers are the fundamental building blocks of IT. Merriam-Webster defines a computer as a programmable, usually electronic, device that can store, retrieve and process data.[1] This definition leaves a lot of room for discussion about what exactly a computer is. Section 1 explores the items, components and parts that make up a computer, including the different formats, shapes and form factors of computers.

1.3 Basic Data Concepts

This section describes the different categories of data in computer systems. A computer receives, stores, processes and outputs data. When the data is collected, it can be stored in memory to use later, or may be processed right then.

Topics covered in this subsection include:

1. Data Basics
2. Data Input
3. Data Storage
4. Data Processing
5. Data Output

[1] Merriam-Webster, "Computer," https://www.merriam-webster.com/dictionary/computer#other-words

1.3.1 Data Basics

Computers use binary language of 1s and 0s. This section explains how a computer uses binary language to receive, process and store data.

Humans use decimal numbering which is base 10 and is spoken and written as 0 to 9. Decimal numbering is the commonly-used base 10 numbering system taught in elementary school math; every digit has 10 possible numbers: 0 through 9. Each digit of a number has a place holder for the value of that part of the number. To break down the number 324, it can be expressed as $3x10^2 + 2x10^1 + 4x10^0 = 324$. It can also be written like this $3x100 + 2x10 + 4x1 = 324$. Elementary math might use small tables that show the values of each place or digit (**figure 1.1**).

Figure 1.1—Decimal Numbering

Exponent Value	10^2	10^1	10^0	
Place Value	100s Place	10s Place	1s Place	Total Number
	3	2	4	324
	4	7	8	478

Binary works the same way; however, it is base 2 instead of base 10. The only possible values are 0 and 1. Every number can be still be expressed, but with a series of 0s and 1s instead of the decimal numbers which are more common. With binary digits, the place holders double as they are exponents of 2. So if using the same 324 number from the decimal example, in binary, it is expressed like this: $1x2^8 + 0x2^7 + 1x2^6 + 0x2^5 + 0x2^4 + 0x2^3 + 1x2^2 + 0x2^1 + 0x2^0 = 324$. The same information in the place value table may be easier to visualize (**figure 1.2**).

Figure 1.2—Binary Decimal Numbering

Exponent Value	2^8	2^7	2^6	2^5	2^4	2^3	2^2	2^1	2^0	
Place Value	256s Place	128s Place	64s Place	32s Place	16s Place	8s Place	4s Place	2s Place	1s Place	Total Number
	1	0	1	0	0	0	1	0	0	324
	1	1	1	0	1	1	1	1	0	478

Even for a reasonably small 3-digit decimal number, it takes 9 binary digits to represent it. The number of binary digits continues to quickly outpace decimal digits, growing to an unusable number of digits even more quickly—especially when those digits are being transmitted over networks where loss or errors can easily happen. To help with this issue, computers typically only deal with groups of binary digits which are called bits. A bit is the most basic and the smallest unit of information in computing. Bits are grouped into sets of 8 which is called a byte. Stated differently, a byte is 8 times more than one bit. Typically, in computing, a byte is the smallest unit of measure that is used.

The terms kilobyte and kilobit are prominent with Internet Service Providers (ISP), and most technology devices on the market. The most notable example is between storage (as on a hard drive) and network transfer rates (such as, the speed of a switch port). Storage is typically notated in kilobytes (KB) and network throughput is notated in kilobits per second (Kbps).

Kilo means one thousand, so one thousand bits is a kilobit and one thousand bytes is a kilobyte. With decimal or base 10, kilo means exactly 1000, but with binary 2^{10} is 1024. This difference can become even greater with units beyond just kilo. There is mega, giga, tera, peta and so on. The International System of Units (SI) prefixes are intended to be used only for base 10 numbers. The unit prefixes are defined and controlled by the General Conference on Weights and Measures.

The list of approved SI prefixes is given in the following table (**figure 1.3**):

Figure 1.3—SI Prefixes

Multiples			Submultiples		
Factor	Prefix		Factor	Prefix	
	Name	Symbol		Name	Symbol
10^1	deca	da	10^{-1}	deci	d
10^2	hecto	h	10^{-2}	centi	c
10^3	kilo	k	10^{-3}	milli	m
10^6	mega	M	10^{-6}	micro	µ
10^9	giga	G	10^{-9}	nano	n
10^{12}	tera	T	10^{-12}	pico	p
10^{15}	peta	P	10^{-15}	femto	f
10^{18}	exa	E	10^{-18}	atto	a
10^{21}	zetta	Z	10^{-21}	zepto	z
10^{24}	yotta	Y	10^{-24}	yocto	y

Source: International Electrotechnical Commission (IEC), "SI prefix," http://www.electropedia.org/iev/iev.nsf/display?openform&ievref=112-02-03

There is a similar, but distinct, set of prefixes used with the binary numbering system. To continue with our example, 1 kilobit should not be used to represent 1,024 bits. The standard prefix for binary is 1 kibibit.

Figure 1.4 shows prefixes for binary multiples:

Figure 1.4—Binary Multiples Prefixes

Factor	Value	Prefix	
		Name	Symbol
$(2^{10})^1$	1 024	kibi	Ki
$(2^{10})^2$	1 048 576	mebi	Mi
$(2^{10})^3$	1 073 741 824	gibi	Gi
$(2^{10})^4$	1 099 511 627 776	tebi	Ti
$(2^{10})^5$	1 125 899 906 842 624	pebi	Pi
$(2^{10})^6$	1 152 921 504 606 846 976	exbi	Ei
$(2^{10})^7$	1 180 591 620 717 411 303 424	zebi	Zi
$(2^{10})^8$	1 208 925 819 614 629 174 706 176	yobi	Yi

Source: International Electrotechnical Commission (IEC), "binary refix," http://www.electropedia.org/iev/iev.nsf/display?openform&ievref=112-01-27

After understanding how bits and bytes are accounted for, there is still the issue of very large binary numbers which contain too much data to use efficiently and without errors. This is the main reason for the use of hexadecimal. Hexadecimal is the base 16 numbering system; hexadecimal allows an entire byte of data to be written in two characters instead of 8 bits.

Most computer users today have probably seen hexadecimal notation, even if they did not know what they were looking at. Hexadecimal is used in defining colors in most graphic-editing programs, so to change the color of a picture and pick a custom color, often the hex value for the color is displayed next to it. Hexadecimal notation uses numbers and letters.

As discussed with base 10 numbering, each digit has ten possible values.

Recall that base 2 numbering systems have two possible values. In base 16 notation, each digit has 16 possible values. Since there are only 10 numbers in the Arabic numeral system, another symbol must be used, so hexadecimal notation also uses letters for readability. Hexadecimal digits more closely resemble base 10 numbering. **Figure 1.5** shows the corresponding values in each numbering system.

Figure 1.5—Hexadecimal Value to Binary Value

Decimal Value	HEX Value	Binary Value
0	0	0000
1	1	0001
2	2	0010
3	3	0011
4	4	0100
5	5	0101
6	6	0110
7	7	0111
8	8	1000
9	9	1001
10	A	1010
11	B	1011
12	C	1100
13	D	1101
14	E	1110
15	F	1111

The table visually shows that Hexadecimal is easier to read than binary, but it also depicts the second major factor for using hexadecimal: higher information density. Two hexadecimal digits can express any number from 0 to 255, which would require 8 digits in binary. To quickly figure out which number is bigger in a comparison, looking at a8 and d6 is easier than 10101000 and 11010110. A significant factor for implementation is how easy it is to read, understand and compare different numbers. Again, hexadecimal uses the digits 0–9 and letters a-f. It is simple to look at the numbers 4 and 7 and know immediately that 7 is greater than 4. It is intuitive to look at "a" and "c" and know just as quickly that "c" comes after "a" and is therefore a higher value.

So why not use base 10—decimal—our everyday counting system? First, computers use binary to process and calculate data. Binary is the base language of computers for the foreseeable future, which means that any other numbering systems used for readability, density, transmission and conversion need to be easy to calculate back and forth to binary. The biggest problem with using decimal is that its digits are not common denominators or multiples with the digits of binary. This is easier to visualize on a number line (**figure 1.6**).

Figure 1.6—Comparison of Binary, Decimal and Hexadecimal Digits

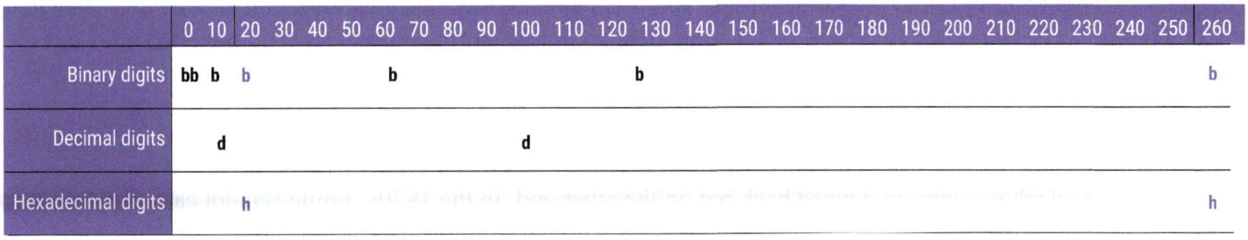

Visually, binary and decimal do not overlap mathematically, but hexadecimal and binary do at the decimal values of 16 and 256. Those points of alignment are equal to 4 binary digits, or half a byte. Half a byte, or 4 binary digits, is sometimes called a nibble.

The example given above for comparing numbers shows this exactly. There are two hexadecimal digits that represent the binary byte. One hexadecimal digit can represent four binary digits. Decimal does not work this way. The multiples for decimal are 10, 100, 1000 and so on, and these multiples never line up exactly with binary numbers. So, any numbering system that gets used should have a base that is a multiple of 2. What about higher bases such as base 64, 128 or 256?

The following table describes how base 64 comes close to being a usable system, since binary and base 64 also align with each other. However, to make base 64 usable, there must be a reference character for every base value—64 different values. Using the entire English alphabet and Arabic numbers there are 36 characters, and using capital and non-capital letters allows 52, plus 10 numbers for a total of 62. To get 64 symbols the "+" and "/" are used. In **figure 1.7**, all the symbols are displayed with their values.

Figure 1.7—Base 64 Values and Associated Characters

Value	Character	Value	Character	Value	Character	Value	Character
0	A	16	Q	32	g	48	w
1	B	17	R	33	h	49	x
2	C	18	S	34	i	50	y
3	D	19	T	35	j	51	z
4	E	20	U	36	k	52	0
5	F	21	V	37	l	53	1
6	G	22	W	38	m	54	2
7	H	23	X	39	n	55	3
8	I	24	Y	40	o	56	4
9	J	25	Z	41	p	57	5
10	K	26	a	42	q	58	6
11	L	27	b	43	r	59	7
12	M	28	c	44	s	60	8
13	N	29	d	45	t	61	9
14	O	30	e	46	u	62	+
15	P	31	f	47	v	63	/

Using base 128 or base 256, both multiples of 2, would need symbols like "+" and "/" to fill out all the value spots. That would mean using all the available keyboard symbols and a few more that are not on the keyboard. Then, unlike knowing the order of letters and numbers, it would require rote memorization to remember the random values of these symbols to read and compare them to one another. This is not a fun task.

So, why bytes? Why 8 bits and not some other amount? It is interesting to consider an encoding system that would take data in binary format from a computer and send it or compare it easily. The next question is, how does data get into a computer, to be processed as binary data. People read and speak in letters, numbers and punctuation. Binary requires a system that takes all commonly used symbols and translates them into something the computer can use.

The need to convert language to letters and numbers started even before computers and computing devices. Before computers, there were teletype machines. Teletypes had to convert letters into signals and sounds that could be transmitted over telephone wires and typed back out on the other end. In the 1870s, Emile Baudot came up with the

Baudot code, a 5-bit mapping for use on teletypes and manual keyboards. 5-bit codes could only support letters and some punctuation.[2] The evolution continued.

As computers became more complex, like at IBM in the 1960s, 6-bit architectures such as binary-coded decimal (BCD) were developed. A 6-bit encoding system allows for 64 values/characters, which means more numbers, special characters or upper and lowercase letters. As in the table above, using uppercase letters, lowercase letters and numbers does not allow for more than two special characters. So 6-bit systems did not last very long.

With the limitations of earlier systems, it made sense to move on to a 7-bit system. This gave room for all characters in uppercase letters, lowercase letters, numbers and special characters. It also left room for extra codes, most often called system codes which make up the first 32 spots of a 0–31 character set.

The system codes could be things like "Start of text", "end of text" or "acknowledge". This was very useful and was formalized in 1963 as the 7-bit ASCII character set, shown in **figure 1.8**. There is also an extended version of ASCII that is shown in **figure 1.9**.

Figure 1.8—7-Bit ASCII Character Set

Dec	Hex	Oct	Character
0	0	000	NUL (null)
1	1	001	SOH (start of heading)
2	2	002	STX (start of text)
3	3	003	ETX (end of text)
4	4	004	EOT (end of transmission)
5	5	005	ENQ (enquiry)
6	6	006	ACK (acknowledge)
7	7	007	BEL (bell)
8	8	010	BS (backspace)
9	9	011	TAB (horizontal tab)
10	A	012	LF (NL line feed, new line)
11	B	013	VT (vertical tab)
12	C	014	FF (NP form feed, new line)
13	D	015	CR (carriage return)
14	E	016	SO (shift out)
15	F	017	SI (shift in)
16	10	020	DLE (data link escape)
17	11	021	DC1 (device control 1)
18	12	022	DC2 (device control 2)
19	13	023	DC3 (device control 3)
20	14	024	DC4 (device control 4)
21	15	025	NAK (negative acknowledgment)
22	16	026	SYN (synchronous idle)
23	17	027	ETB (end of trans. block)
24	18	030	CAN (cancel)
25	19	031	EM (end of medium)
26	1A	032	SUB (substitute)
27	1B	033	ESC (escape)
28	1C	034	FS (file separator)
29	1D	035	GS (group separator)
30	1E	036	RS (record separator)
31	1F	037	US (unit separator)

[2] Calvert, J.B.; "Teletypewriter," 8 August 2002, http://mysite.du.edu/~jcalvert/tel/teletype.htm

Figure 1.8—7-Bit ASCII Character Set *(cont.)*

Dec	Hex	Oct	HTML	Character
32	20	040		Space
33	21	041	!	!
34	22	042	"	"
35	23	043	#	#
36	24	044	$	$
37	25	045	%	%
38	26	046	&	&
39	27	047	'	'
40	28	050	(	(
41	29	051	)	)
42	2A	052	*	*
43	2B	053	+	+
44	2C	054	,	,
45	2D	055	-	-
46	2E	056	.	.
47	2F	057	/	/
48	30	060	0	0
49	31	061	1	1
50	32	062	2	2
51	33	063	3	3
52	34	064	4	4
53	35	065	5	5
54	36	066	6	6
55	37	067	7	7
56	38	070	8	8
57	39	071	9	9
58	3A	072	:	:
59	3B	073	;	;
60	3C	074	<	<
61	3D	075	=	=
62	3E	076	>	>
63	3F	077	?	?
64	40	100	@	@
65	41	101	A	A
66	42	102	B	B
67	43	103	C	C
68	44	104	D	D
69	45	105	E	E
70	46	106	F	F
71	47	107	G	G
72	48	110	H	H
73	49	111	I	I
74	4A	112	J	J
75	4B	113	K	K
76	4C	114	L	L

Figure 1.8—7-Bit ASCII Character Set *(cont.)*

Dec	Hex	Oct	HTML	Character
77	4D	115	M	M
78	4E	116	N	N
79	4F	117	O	O
80	50	120	P	P
81	51	121	Q	Q
82	52	122	R	R
83	53	123	S	S
84	54	124	T	T
85	55	125	U	U
86	56	126	V	V
87	57	127	W	W
88	58	130	X	X
89	59	131	Y	Y
90	5A	132	Z	Z
91	5B	133	[	[
92	5C	134	\	\
93	5D	135	]	]
94	5E	136	^	^
95	5F	137	_	_
96	60	140	`	`
97	61	141	a	a
98	62	142	b	b
99	63	143	c	c
100	64	144	d	d
101	65	145	e	e
102	66	146	f	f
103	67	147	g	g
104	68	150	h	h
105	69	151	i	i
106	6A	152	j	j
107	6B	153	k	k
108	6C	154	l	l
109	6D	155	m	m
110	6E	156	n	n
111	6F	157	o	o
112	70	160	p	p
113	71	161	q	q
114	72	162	r	r
115	73	163	s	s
116	74	164	t	t
117	75	165	u	u
118	76	166	v	v
119	77	167	w	w
120	78	170	x	x
121	79	171	y	y
122	7A	172	z	z
123	7B	173	{	{
124	7C	174			|

Figure 1.8—7-Bit ASCII Character Set *(cont.)*				
Dec.	**Hex**	**Oct**	**HTML**	**Character**
125	7D	175	}	}
126	7E	176	~	~
127	7F	177	&#127;	DEL
Source: American National Standards Institute, https://www.ansi.org/				

Figure 1.9—Extended ASCII Character Set (Microsoft Windows -1252)				
Dec.	**Hex**	**Oct**	**HTML**	**Character**
128	80	200	€	€
129	81	201		[unprintable character]
130	82	202	‚	‚
131	83	203	ƒ	ƒ
132	84	204	„	„
133	85	205	…	…
134	86	206	†	†
135	87	207	‡	‡
136	88	210	‡	ˆ
137	89	211	‰	‰
138	8A	212	Š	Ŝ
139	8B	213	‹	‹
140	8C	214	Œ	Œ
141	8D	215		[unprintable character]
142	8E	216	Ž	Ž
143	8F	217		[unprintable character]
144	90	220		[unprintable character]
145	91	221	‘	‘
146	92	222	’	’
147	93	223	“	“
148	94	224	”	”
149	95	225	•	•
150	96	226	–	–
151	97	227	—	—
152	98	230	˜	˜
153	99	231	™	™
154	9A	232	š	ŝ
155	9B	233	›	›
156	9C	234	œ	œ
157	9D	235		[unprintable character]
158	9E	236	ž	ž
159	9F	237	Ÿ	Ÿ
160	A0	240		[no-break space]
161	A1	241	¡	¡
162	A2	242	¢	¢
163	A3	243	£	£
164	A4	244	¤	¤
165	A5	245	¥	¥
166	A6	246	¦	¦

Figure 1.9—Extended ASCII Character Set (Microsoft Windows -1252) *(cont.)*

Dec.	Hex	Oct	HTML	Character
167	A7	247	§	§
168	A8	250	¨	¨
169	A9	251	©	©
170	AA	252	ª	ª
171	AB	253	«	«
172	AC	254	¬	¬
173	AD	255		[soft hyphen]
174	AE	256	®	®
175	AF	257	¯	¯
176	B0	260	°	°
177	B1	261	±	±
178	B2	262	²	²
179	B3	263	³	³
180	B4	264	´	´
181	B5	265	µ	µ
182	B6	266	¶	¶
183	B7	267	·	·
184	B8	270	¸	¸
185	B9	271	¹	¹
186	BA	272	º	º
187	BB	273	»	»
188	BC	274	¼	¼
189	BD	275	½	½
190	BE	276	¾	¾
191	BF	277	¿	¿
192	C0	300	À	À
193	C1	301	Á	Á
194	C2	302	Â	Â
195	C3	303	Ã	Ã
196	C4	304	Ä	Ä
197	C5	305	Å	Å
198	C6	306	Æ	Æ
199	C7	307	Ç	Ç
200	C8	310	È	È
201	C9	311	É	É
202	CA	312	Ê	Ê
203	CB	313	Ë	Ë
204	CC	314	Ì	Ì
205	CD	315	Í	Í
206	CE	316	Î	Î
207	CF	317	Ï	Ï
208	D0	320	Ð	Ð
209	D1	321	Ñ	Ñ
210	D2	322	Ò	Ò
211	D3	323	Ó	Ó
212	D4	324	Ô	Ô
213	D5	325	Õ	Õ

Figure 1.9—Extended ASCII Character Set (Microsoft Windows -1252) *(cont.)*

Dec.	Hex	Oct	HTML	Character
214	D6	326	Ö	Ö
215	D7	327	×	×
216	D8	330	Ø	Ø
217	D9	331	Ù	Ù
218	DA	332	Ú	Ú
219	DB	333	Û	Û
220	DC	334	Ü	Ü
221	DD	335	Ý	Ý
222	DE	336	Þ	Þ
223	DF	337	ß	ß
224	E0	340	à	à
225	E1	341	á	á
226	E2	342	â	â
227	E3	343	ã	ä
228	E4	344	ä	ä
229	E5	345	å	å
230	E6	346	æ	æ
231	E7	347	ç	ç
232	E8	350	è	è
233	E9	351	é	é
234	EA	352	ê	ê
235	EB	353	ë	ë
236	EC	354	ì	ì
237	ED	355	í	í
238	EE	356	î	î
239	EF	357	ï	ï
240	F0	360	ð	ð
241	F1	361	ñ	ñ
242	F2	362	ò	ò
243	F3	363	ó	ó
244	F4	364	ô	ô
245	F5	365	õ	õ
246	F6	366	ö	ö
247	F7	367	÷	÷
248	F8	370	ø	ø
249	F9	371	ù	ù
250	FA	372	ú	ú
251	FB	373	û	û
252	FC	374	ü	ü
253	FD	375	ý	ý
254	FE	376	þ	þ
255	FF	377	ÿ	ÿ

Source: American National Standards Institute, https://www.ansi.org/

Even though 7-bit systems have most necessary characters, a power of 2 is still better, so 8 bits makes more sense than 7 bits. The extra, or 8th, bit is often used as a "parity bit." The parity bit is typically the first, or leading, bit of the character string. It usually indicates if the number of 1s in the following 7 bits is even or odd. As computers and transmissions at that time were very error prone, this parity could be used for error-detection.

The next section describes how the data are input to a computer for use.

1.3.2 Data Input

Almost any data that a computer receives can be looked at as input data. For example when a person sees or hears about something happening, it is similar to a computer being given some sort of input data by an input device.

Keyboard

A keyboard has long been the most common input device. A keyboard is very simple to use and understand which helps to quickly enter data into the computer. The keyboard is laid out much like a traditional typewriter, but computers have additional keys for performing functions not available on typewriters and systems of the past (**figure 1.10**).

Figure 1.10—Keyboard

Source: Onebluelight; Getty Images

Keyboards traditionally came in a couple of sizes: 84 keys or 101/102 keys, but now keyboards have 104 keys or 108 keys. Many have special keys for the operating system (Windows or Mac), the Internet and even for gaming. Gaming keyboards might have color coded keys, lights and special layouts of keys to make playing games easier.

Mouse

The mouse has long been the most common pointing device. It is a cursor-control device, which was first referred to as an X-Y pointing device at the time of its invention. It is typically a small palm-size box or oval, that has a spherical ball on the bottom (**figure 1.11**). The ball rolls over a surface, such as a desk, which spins small wheel sensors. That movement is translated into corresponding signals to the CPU which shows up on the screen as cursor movement. This allows the mouse to move the cursor much faster that with a keyboard.

Most newer mice no longer have a ball in them that moves—the movement now is sensed and calculated by lasers in the base of the mouse. Even before lasers , the mouse had other upgrades. It started off with just two buttons called the left and the right buttons. They allowed for simple "clicking" to make a selection or move objects on the screen. A wheel was introduced between the buttons to allow for scrolling, which made large documents and Internet pages easier to navigate.

Figure 1.11—Mouse

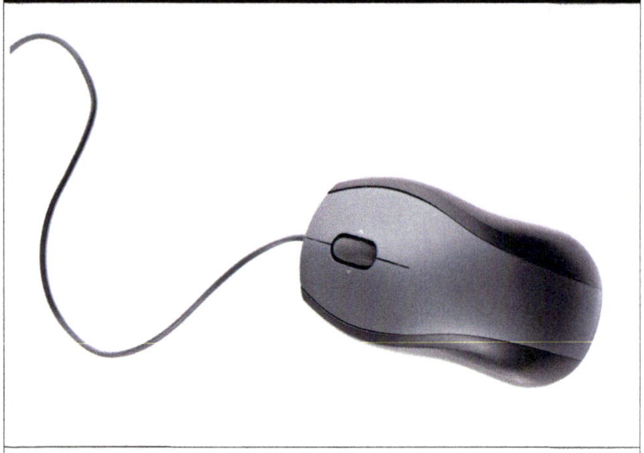

Source: +Rrocio; Getty Images

Joystick

A joystick is also a pointing device. Similar to a mouse, it is used to move the cursor position on a monitor screen. It has a moveable ball that sits in a socket. Attached to the ball, in the socket, is a vertical "stick". Sometimes, on the top of that stick is another ball, as shown in **figure 1.12**. But other joysticks just have small protrusions, or flat tops to act as a thumb rest. At one time, the main use for a joystick was in Computer Aided Design (CAD) and similar software, but currently, joysticks are primarily used for playing computer games.

Figure 1.12—Joystick

Source: Robstyle; Getty Images

Trackball

In the evolution of input pointing devices, the trackball was created for a new type of computer system—notebook or laptop computers. The trackball is set up opposite a mouse, similarly to a joystick, where the ball is half inserted into the top of the device's body. The pointer moves by moving fingers on the ball and rolling the ball around. (**figure 1.13**).

Since only the ball moves, and not the whole device, a trackball does not need a large movement space like a mouse. A trackball comes in various styles and shapes similar to other pointing devices.

Figure 1.13—Trackball

Source: Stryjek4; Getty Images

Stylus, Touch Pen and Light Pen

The first stylus pens (**figure 1.14**) were nothing more than a pen-shaped device which had typically a plastic tip that would not scratch or damage a monitor or screen. This simple pen did not do any of the work—the screen had to be a touch-sensitive screen. These first touch screens had layers in them that, when pressed together, made a circuit that would register the placement of the touch pen.

A stylus, more commonly seen today, has a tip that is made of a specific type of plastic. This plastic-type allows the stylus to have a capacitive charge effect on newer capacitive touch screens. These capacitive touch screens have an electrical charge going through them, so that when the capacitive pen touches the screen, it changes the charge in that spot, thereby registering a touch.

A light pen (**figure 1.15**) is a computer-input device which looks like a normal pen, but is light-sensitive and must be used with a cathode-ray tube (CRT) display. In the past few years, most CRT monitors have been replaced by newer types of monitors. CRT monitors are only used today for a few specific industrial applications, where the monitor's ability to be placed in a protective enclosure and heat and magnetism resistance is required. The Output Devices section describes the differences in the various types of monitors and displays. The light pen points to displayed objects or draws on the screen, similar to a finger on a touchscreen.

Figure 1.14—Stylus (or Touch Pen)

Source: Part; Getty Images

Figure 1.15—Light Pen

Source: Luismolina; Getty Images

Scanners

A traditional "bed" or layout scanner works like a photocopier (**figure 1.16**). Some models have paper feeders that allow for a stack of papers, photos or business cards to be scanned and input to the computer. This type of device became popular in homes when people wanted to convert paper-based photos to the computer to send in email or use in digital documents.

Figure 1.16—Scanner

Source: Thomas Northcut; Getty Images

An optical character reader (OCR) is an input device used to read and convert printed text from a static image into the computer (**figure 1.17**). At one time, this device existed on its own. Now, most scanners do this same function and the text is converted and translated using software to identify and understand it.

Figure 1.17—Optical Character Reader

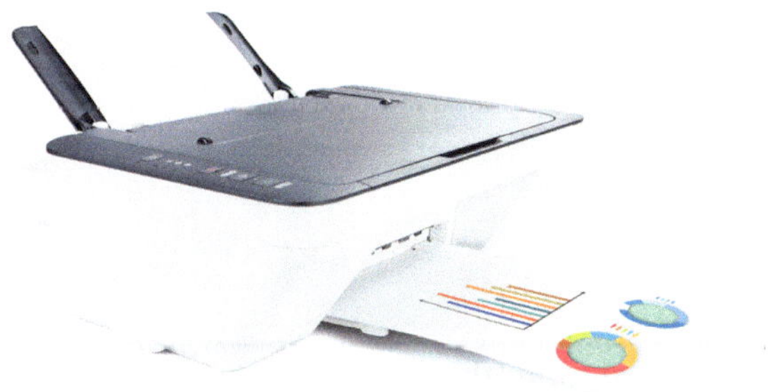

Source: Turk_stock_photographer; Getty Images

Barcode readers are used for reading and scanning barcodes. Barcodes are used to label just about everything—groceries in the store, books or shipping labels. Cell phones have apps that use the camera on the phone to scan barcodes. When simple barcodes (vertical white and black lines of varying width) no longer store enough data, other types of codes were developed, such as quick response (QR) codes (**figure 1.18**).

Figure 1.18—Sample QR Code

Source: Youngvet; Getty Images

Digital Converter

Sometimes referred to as a digitizer, a digital converter is an input device which converts analog to digital information (**figure 1.19**). There are many different devices and items that are marketed under this category. Some digitizers can take a signal from a television or camera and convert it into digital data to be used and stored on a computer. Some of these devices are like tablets, such as below:

Figure 1.19—Digital Converter

Source: Nirut Sangkea/EyeEm; Getty Images

These tablets convert written or drawn things into digital data.

Microphone

A microphone is an input device used to convert sound waves into digital form (**figure 1.20**). Not only do these devices allow us to add voice data to presentations and videos, but they are also built into every-day devices. Digital phones in the home and office would not exist without this ability. Cell phones, which are all digital, would not exist without it.

Figure 1.20—Microphone

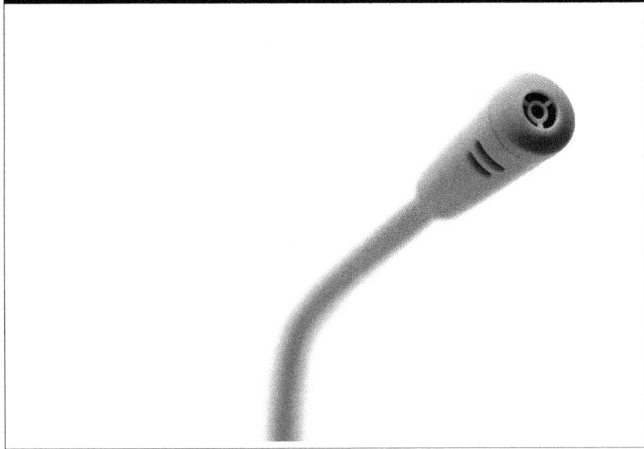

Source: Jupiterimages; Getty Images

Voice over Internet Protocol

Voice over Internet Protocol (VoIP) applications and systems rely on a computer system's ability to sample analog sounds to measure the analog sound. After sampling sounds, a VoIP device converts those samples into digital strings of data, transfers them over a network to another computer or device which modulates the strings of data back into sound waves that can be heard on the other end.

Voice and video calls on the computer are not the only use for microphones. Today, there is a whole new industry comprised of content creators. They create informative, educational, philosophical and entertainment shows of all kinds and post them to online forums. The shows take many forms, including podcasts and video blogs (VLOGs). This content is produced in varying levels of quality and equipment. For example, a podcast could be created with the built-in microphone on a camera or studio-level, boom-mounted microphones.

1.3.3 Data Storage

Storage on a computer is called memory, and is used to hold data and instructions. Memory can be categorized into four types—main, internal, secondary and tertiary. These four main types are based on their order, or location, as data is processed. Scarcity, cost and intended use are all important criteria to look at when determining the memory/storage requirement of any computer system. **Figure 1.21** shows the computer memory hierarchy.

Figure 1.21—Computer Memory Hierarchy

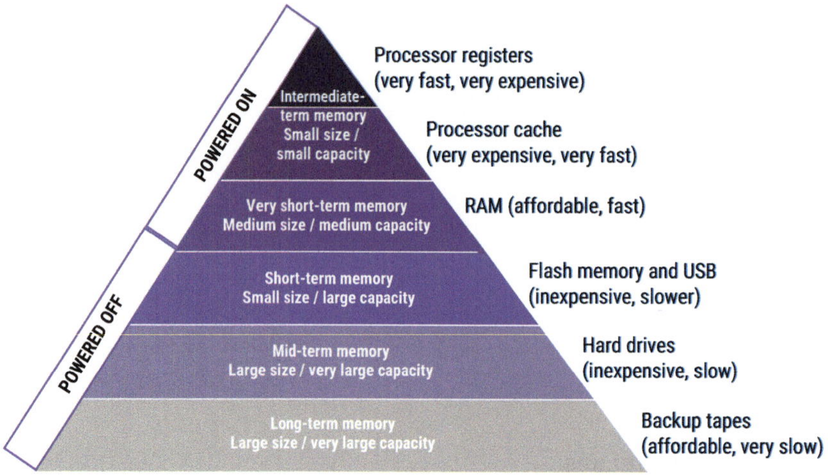

At the top of the model is the fastest and most expensive memory. At the bottom is the slowest and least expensive types of memory. Of the six layers in the model, this section defines four main categories, and within those categories, some differences in where they reside and how the computer CPU can access them. The four categories are:

- Internal Memory
- Main Memory
- Secondary Storage
- Tertiary Storage

The first two categories, internal and main, are also known as direct access memory, and make up the top three layers of the memory hierarchy model. Direct access means the CPU can read and write directly to them. The last two categories, secondary and tertiary, fall into the bottom three layers of the memory model, and are also known as external memory. This means that the CPU cannot directly read and write to them. External also means that they don't have to be connected or accessible all the time for the system to run, operate and do calculations. Because that is the case, these types are also classified as being online, nearline or offline.

Internal Memory

The top two levels of the computer memory hierarchy model represent what is typically called internal memory. Internal memory is made of up processor registers and cache memory. Even within processor registers, there are different types of registers used for different types of data. The different types are not important right now, but these registers provide the fastest access to memory by the CPU. This is as close to instant as it gets in current hardware architecture and can happen in a single cycle of the CPU.

Cache memory is semiconductor-based memory and the highest-speed memory of its type. Cache memory holds the most often-used parts of a program or instruction set for the CPU. It is the go-between for the CPU and main memory. Main memory is the next layer down in the model. The operating system of the computer copies the program and instructions from lower levels in the hierarchy model) of memory into cache. Cache is the memory that the CPU can directly access to complete calculations. Cache memory is broken up into multiple sublevels of memory known as level 2 or level 3 cache. More cache is better, but more cache costs more money. There is typically fewer lower-numbered cache types, but they have fast access time to the CPU. Even the slowest cache today has an access time of 40 to 100 GiB/second, and the fastest can be over 700 GiB/second.

Main Memory

The third level in the computer memory hierarchy model is called main memory or primary memory (**figure 1.22**) . Main memory is the primary data storage location for data in use, which includes only the instructions that are currently being processed. Main memory can consist of random access memory (RAM) which is volatile and read-only memory (ROM) which is non-volatile. Volatile memory means that if power is lost to the memory modules, then whatever data was being stored is lost. Non-volatile memory does not have this characteristic—data written to it survives power loss. Primary memory includes both RAM and ROM which are accessed directly by the CPU, although main memory is frequently just referred to as RAM. RAM is a fairly typical metric for measuring the speed and quality of a computer system. This is due to the limited amount of RAM memory, as it is costly, semiconductor-based memory.

Figure 1.22—RAM Memory Chip

Source: Peterhowell; Getty Images

Secondary Memory

Secondary memory makes up the fourth and fifth layers of the model—it is also called external memory or non-volatile memory. If main memory is volatile, data would be lost; in secondary memory data will not be lost due to a power outage. Further down the computer memory hierarchy, external memory describes the CPU's inability to access these memory types directly, and any type lower in the model. The operating system accesses this type of memory and uses input-output (I/O) routines to control and move the data. Secondary memory can be flash-based, as in layer 4 of the model. This might be small flash drives or solid state drives (SSD). The fifth layer of the model includes hard disk drives (HDD) (**figure 1.23**), CD-ROM, DVD, Blu-ray, and the like. Many of the optical storage mediums, such as CD-ROM and DVD formats, are mostly obsolete in their physical use, but these technologies have brought about much research in the field. In the future, it may be possible to see fifth-layer storage devices as electrical, magnetic and optical in design.

Figure 1.23—Hard disk drive

Source: Craig Russell/EyeEm; Getty Images

Tertiary Memory

Tertiary memory is the sixth and last level of the model. It makes up the mass storage of backup drives and backup tapes. This memory does not require power and is not required to be physically connected to the system. This storage may require human intervention to replace and load tapes onto which data can be copied.

Accessibility is often talked about with the lower, external types of storage. This classification has to do with the longevity of the storage medium. A failed HDD can happen as many types of mass storage devices are mechanical. They have magnetic or electric metal platters that are spinning quickly so that data can be written and read to them. This means that they can break down and fail. To keep this failure from happening, other devices have been created, and additional strategies are used to keep important data safe. Other devices and strategies typically will have either a higher actual cost or not be readily accessible.

Online storage is connected to the computer system so that I/O is immediately available. This type of storage goes to disks and tapes that are connected and always spinning.

Nearline storage is very close to online, but instead of keeping the disks spinning and allowing ongoing wear on the system, this type of system can include spin-down of the disks or removal of the tape cassettes. Systems that spin the disks down include massive array of idle disk (MAID) or robotic tape banks where a robotic mechanism can load and unload disks. For nearline storage, there is no human interaction needed.

Offline storage means that not only do the disks and tapes spin down, but they are removed from the system. The media needs human interaction to unload and load the media. This is a good practice, especially to keep backups in a different physical location. If the primary location is destroyed, then the media that was moved to a secondary location will survive with the data.

1.3.4 Data Processing

Data taken in through various inputs is stored and moved to various types and parts of memory. The true work of a computer is data processing. Data processing is the system of collecting and manipulating data to create useful output. The term data processing is fairly new, but the act of data processing has been done for thousands of years. While not as common as the common personal computer, an abacus has been used for calculations and data processing for about 4,500 years.

The modern computer is semiconductor-based and has the ability to do billions of calculations per second. It does that work in the central processing unit (CPU). A CPU is a microprocessor, which is a multipurpose device. It uses registers, and it takes input and output data in binary form. A register is an accessible location to the processor that may hold instructions, storage address or other data such bit sequence or other characters.

All computers, no matter what their function is, must have some type of CPU. A CPU has multiple components within it. The control unit (CU) and the arithmetic/logic unit (ALU) are two key parts. Each does a specific part of the work (**figure 1.24**).

Figure 1.24—CPU Components

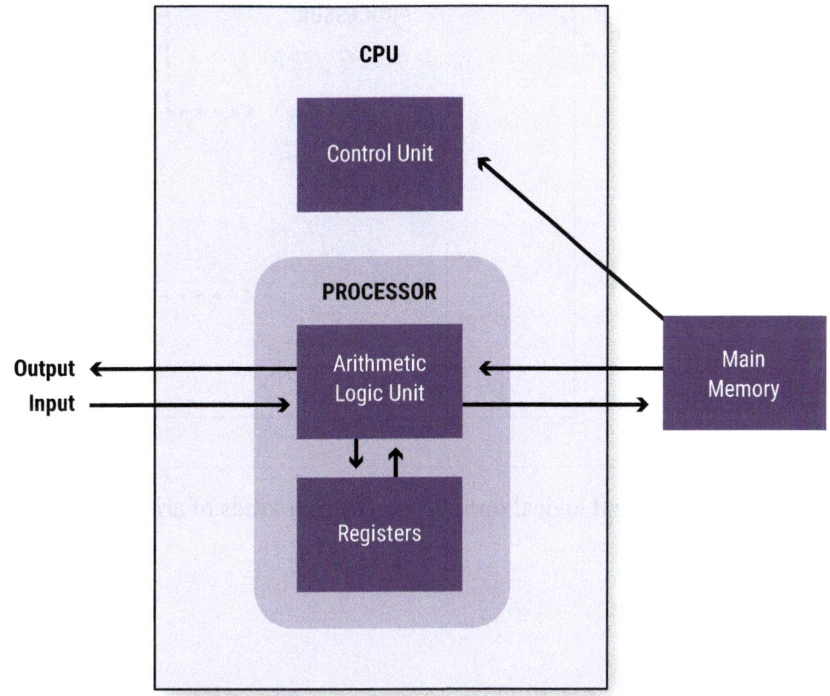

The CU and the ALU are built into the CPU. The CU of the CPU directs the functions of the whole computer system. The figure above shows the data paths from each of the parts. Additional lines from the control unit (CU) to every other piece, with dotted lines as in the figure below, are the control paths. The control paths are the CU telling everything what the timing is, and how to carry out and execute the instructions of a program (**figure 1.25**).

Figure 1.25—CPU Control Unit

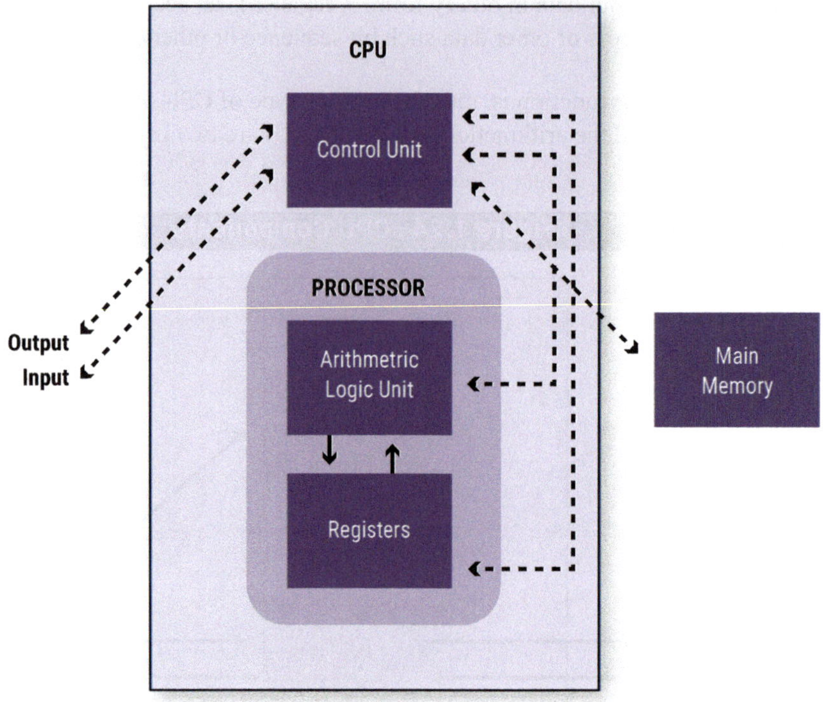

The ALU executes all arithmetic and logical operations. The four kinds of arithmetic operations are:

- Addition
- Subtraction
- Multiplication
- Division

The logic part of the ALU performs logical operations. A logical operation is basically a comparison. When a programmer writes a program, they establish parameters for comparisons. The ALU can compare numbers, letters or special characters. The computer takes action based on the output of the compared data. This is why it is difficult to remember symbols and special characters with a complicated perceived-value at scale. A programmer and someone using the program must know the values of all things being compared. There are millions of comparisons that happen in computer software constantly, including simple things such as quantifying an inventory or checking for new email. These logical operations can test for three conditions: equal to, less than, or greater than.

- **Equal-to** looks at two values to determine if they are equal. If there are 30 apples in one box, are there also 30 apples in another box?
- **Less-than** compares two values to determine if one is less than the other. A banking program may do this with amounts in accounts. An action may happen, such as transferring money from savings to checking, if the amount in a checking account is less than a pending purchase.
- **Greater-than** is focused on identifying values greater than the other. If a system controls the climate of a building and the thermostat is set to 73 degrees, when the temperature rises above 73, the system will turn on the air conditioning.

The computer can test for more than one condition at a time. In fact, there are typically six logical relationships:

- Equal to
- Less than
- Greater than
- Less than or equal to
- Greater than or equal to
- Not equal

Some special characters have specific definitions and are used in the actual programming of the logic. For example, the symbols used for comparison, called relational operators, are the common and expected math symbols. The equal sign "=," the less-than symbol "<" and the greater-than symbol ">."

One of the most interesting things about computers is that they can only execute a single instruction at a time. It is amazing how quickly they can run the single instruction. Most computers today can run billions of instructions per second. Many factors go into how many and how fast instructions can be run.

To execute an instruction, the data and instructions for the program must be moved into memory. That information comes from an input device or secondary storage. When the data and instruction are in the memory, the CPU completes these four basic steps, and will do it for each instruction:

1. The CU "fetches" the instruction from memory.
2. The CU interprets the instruction and directs the required data be moved from memory to the ALU.
3. The ALU runs the arithmetic or logical instruction.
4. The ALU stores the result of this operation in memory or in a register.

Every CPU is essentially a synchronous device, meaning it does one thing at a time on a fixed time schedule. The CPU has an internal clock that creates time pulses at a fixed rate which keeps all the sub-parts (CU, ALU, memory, input, output) synchronized and controlling when operations will happen. This clock sets the machine-cycle, or the clock cycle. For a CPU that is set at 4.0 GHz, the 4.0 GHz describes the clock cycle. This means that the CPU has 4 billion cycles per second, or the CPU can run up to 4 billion instructions in a second.

1.3.5 Data Output

At this last step, the complete flow of data becomes clear: starting by collecting or inputting the data, then storing it, processing it and storing it again. Finally, it will be output that is seen, read and heard. Following are descriptions of different types of common and useful output devices.

Monitors

Monitors are the main output device of a computer and are sometimes also referred to as screens, displays or televisions. Any device that has the ability to be connected to a computer, so that the computer's output stream can be displayed, is an output viewing device. A projector connected to a computer is also an output viewing device.

Most displays form images from very small dots, called pixels. The pixels are arranged in a rectangular pattern that fills the viewable front panel of the monitor. The sharpness and clarity of the output stream depends upon the number of pixels. The smaller each pixel is, the better the image clarity or resolution. It takes more than one illuminated pixel to form a whole character.

Cathode-Ray Tube (CRT) monitors (**figure 1.26**) are the oldest type of monitor covered here, and most been retired by now.

Figure 1.26—CRT Monitor

Source: Photology1971; Getty Images

Most CRT screens are divided into rectangular arrays of character boxes. A character box is a single location on the screen where a single character can be placed. Most of these screens display about 80 characters across and 25 lines of characters up and down.

CRT monitors are large in size, so they take up a lot of space and are heavy to lift and move. They also consume a lot of electricity. For these reasons, CRT monitors are rarely used anymore.

Flat-panel displays (**figure 1.27**) are output devices that are significantly thinner in size, weigh less and use less power compared to a CRT. They can be hung on a wall or worn in a watch. Flat-panel displays are used in car displays, calculators, video games, laptop computers, phones and tablets.

Figure 1.27—Flat Panel Display

Source: Ashwin82; Getty Images

Flat-panel displays can generally be divided into two categories: emissive and non-emissive displays. Emissive displays convert electrical energy into light. Emissive display examples include plasma panels, light-emitting diodes

(LED), and organic LED (OLED) displays. Non-emissive displays, such as liquid-crystal displays (LCDs), use optical effects to convert sunlight or light from other sources into graphics patterns.

Speakers

Another popular output device is a speaker. Opposite of a microphone, a computer takes the digitized sounds from music, singing, voice or other types of recording and converts the digital signal into electric signals that can be modulated as output of the speaker.

Speakers are built in to many devices including monitors, phones, cameras and computers. Typically, built-in speakers sacrifice some audio quality for the convenience of being part of their host device. More types of external speakers exist with a wide range of use, quality and price. External speakers are made to connect to computers, phones and tablets to improve the quality of sound in gaming, conference calls, music and movies. External speakers can be grouped together into theater and studio-quality sound systems so that users can produce songs, videos and online content.

Printers

Printers are output devices that print, transcribe, carve, mill, form information onto or out of paper, film, plastic, wood, steel or other physical material.

Impact printers print characters by striking them on the ribbon, which is then pressed on, and in some cases through, the paper (and carbon paper for additional copies). These printers use low-cost consumables. They are very noisy, do not have good quality and are slow.

Character printers are the printers which print one character at a time; they come in a couple of styles: dot matrix and daisy wheel.

The daisy wheel printing head sits on a wheel with pins corresponding to characters, similar to petals on a flower, which is why it is called a daisy wheel printer (**figure 1.28**). Daisy wheel printers are more reliable than dot matrix, and better quality, but they are slower and more expensive.

Figure 1.28—Daisy Wheel Printer

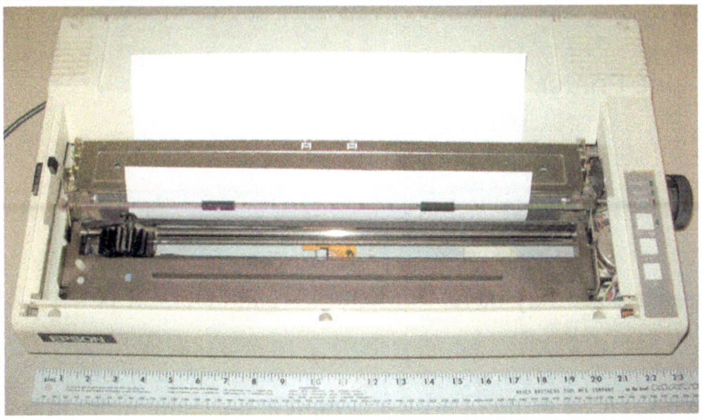

Source: Wikimedia Commons, "Epson Wide Carriage 9-pin printer," https://commons.wikimedia.org/wiki/File:Epson_Wide_Carriage_9-pin_printer_-_with_legal_paper_8.5x14.jpg. This file is licensed under the Creative Commons Attribution-Share Alike 3.0 Unported license

Line printers (**figure 1.29**) print one line at a time.

Figure 1.29—Line Printer

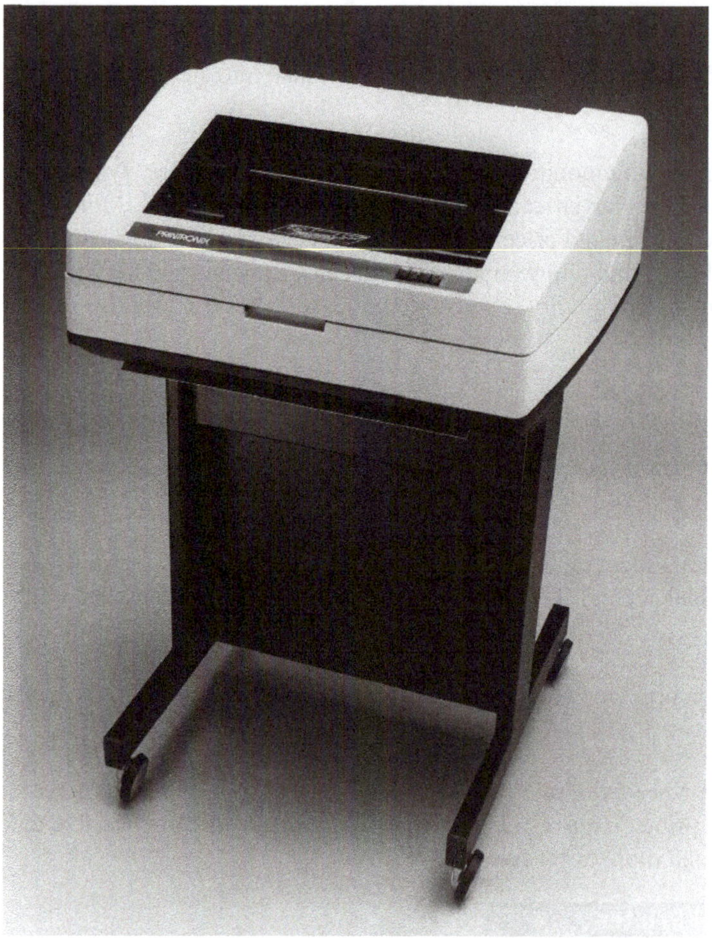

Source: Wikimedia Commons, "Printronix MVP Series Line Printer," https://commons.wikimedia.org/wiki/File:Printronix_MVP_Series_Line_Printer.png. This file is licensed under the Creative Commons Attribution-Share Alike 3.0 Unported license

A chain printer uses a chain of character sets. A standard character set may have 48, 64, or 96 characters. It is possible to change the fonts in this printer, but it is noisy.

Non-impact printers print the characters without using a ribbon. These printers print a complete page at a time, thus they are also called page printers. The two main types of this type of printer are laser printers and inkjet printers. These printers are much faster than impact, and are not nearly as noisy as there is no contact. Non-impact printers provide much higher quality and do not require equipment changes for different fonts, sizes, or languages. They can also print in color.

Laser Printers (**figure 1.30**) use laser lights to produce the dots needed to form the characters printed on a page. Laser printers are much more expensive but have the highest-quality output.

Figure 1.30—Laser Printer

Source: Ryan McVay; Getty Images

Inkjet printers (**figure 1.31**) are also non-impact character printers. They spray small drops of ink onto the paper to create fonts and shapes. Inkjet printers produce high-quality output with presentable features. They can make multiple copies of a page but are slower than laser printers.

Figure 1.31—Inkjet Printer

Source: Thitimon Toyal; Getty Images

3D Printers

3D printing is the most common name of a process called additive manufacturing. In this process, material is added, joined or solidified together to create a three-dimensional object. Material can be liquid or powder that is added layer by layer. The object is "printed" from a computer using computer-aided design (CAD) or 3D-modeling software (**figure 1.32**).

Figure 1.32—3D Printer

Source: Belekekin; Getty Images

CNC Mills and Engravers

Similar to 3D printing, using CAD and modeling software, computer numerical control (CNC) mills and engravers use computer output to create physical objects (**figure 1.33**). They can create objects by drilling, cutting and burning shapes and designs into stock material. The stock material can be metal, plastic, wood or even stone materials. The machines take the code from a CAD design and convert it into a series of numbers that roughly correlate to coordinates of a graph. Using the coordinates, the machine can remove material from a stock to create anything from signs, car wheels and even intricate parts for other machines.

Figure 1.33—CNC Mill and Engraver

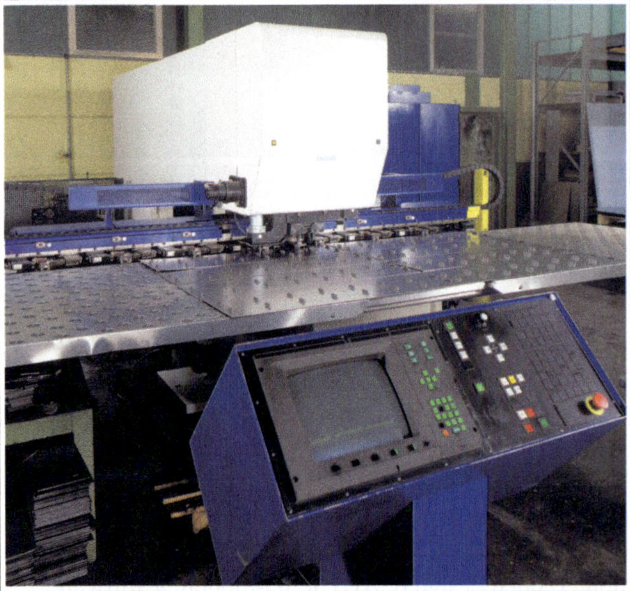

Source: Clu; Getty Images

1.4 Hardware Components of Computing Devices

The name solid-state came about in the 1960s. At the time, electronic devices used vacuum tubes and discrete components. With vacuum tubes, the electronic interaction inside the glass tube was controlled with the current across the electrodes and the electrons in the gas particles moving in the vacuum. Discrete components, being an individual resistor, capacitor, etc., were connected together with wires. The wires and all the individual components allowed electricity to flow and create circuits.

Vacuum tubes operated in a gaseous state. In the 1960s, new semiconductor technology worked based on solid crystalline and could incorporate multiple discrete components into the same device. This brought about the use of the names solid-state electronics and integrated circuit (IC).

1.4.1 Processors and Solid-State Electronics

A fundamental building block of computers and, most electronic devices, is the electronic circuit. An electronic circuit is the combination of multiple components providing some output such as calculations, moving data, or amplification. A typical distinction that is made is the difference between something being electronic and something being electrical. The definition of something being considered electronic is that it must have at least one active component. An electronic circuit falls into one of the following groups: analog circuit, digital circuit, or mixed-signal circuit (a combination of analog circuits and digital circuits). **Figure 1.34** shows a circuit diagram for a common base amplifier. It shows basic discrete components such as resistors and capacitors.

Figure 1.34—Circuit Diagram for Base Amplifier

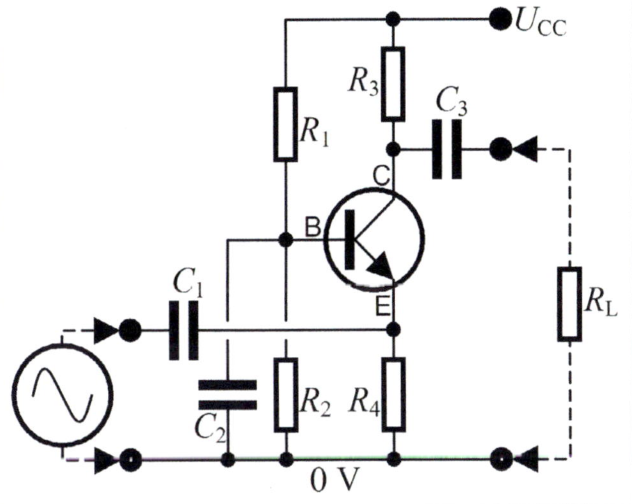

Source: Wikimedia Commons, Common Base amplifier.png, https://commons.wikimedia.org/w/index.php?curid=114152. The figure is licensed under the Creative Commons Attribution-Share Alike 3.0 Unported license.

Circuits

Circuits can be made of discrete components, like capacitors, diodes, inductors, resistors and transistors, which must be connected by conductive wires which electric current can flow through. A more common architecture for a circuit is a printed circuit board (PCB). PCBs connect and mount discrete devices. A PCB is typically made of a laminated substrate with copper layers on or between the substrate layers. The copper layers are etched in a chemical process to leave only the traces and pads needed to solder components that create a finished circuit. Multilayer PCBs are used with layers of copper on one side or both sides of the layer substrate. Additional layers can be created with layers of substrate alternating between layers of copper traces. Layering can be beneficial, especially in the overall size of a completed board, as the space for running the traces between components takes the most real estate. **Figure 1.35** is an example of a PCB.

Figure 1.35—Printed Circuit Board

Source: Peter Dazeley; Getty Images

In an integrated circuit (IC), the components and interconnections are formed on the same substrate, typically a semiconductor. A semiconductor is made primarily of silicon; sometimes with germanium, gallium arsenide and organic semiconductors. Semiconductors work similarly to other circuits by regulating electric current. Electric current consists of electrons or holes, and the electrons move through the silicon material. The requirements for the semiconductor determine what happens in the rest of the circuit. Semiconductors can have many types of transistors and diodes built into them by adding impurities—also called doping. The most widely used semiconductor is the MOSFET (metal-oxide-semiconductor field-effect transistor).

The term IC is also used for devices which have no moving parts. ICs can replace devices with moving parts, such as relays. Instead of a moving-arm electromechanical relay, a solid-state device can be used, just as solid-state drives (SSDs) are replacing spinning hard disk drives (HDDs).

From the very start, when semiconductors had only a few transistors on an IC, people considered how many transistors could be put on a single chip, how small the chip could be, and how much it would cost to produce these new devices. There is a man who is credited for essentially going out on a limb and making a prediction in 1965 about how this growth and density would happen. He was the co-founder of Fairchild Semiconductor and later co-founder of NM Electronics which would become Intel Corporation. Gordon Moore, was asked by Electronics magazine to make the prediction. Moore noted that the number of components (transistors, resistors, diodes or capacitors) in an IC had been close to doubling every year. Moore did some calculations based on how they had been developing ICs up to that point, and concluded that the doubling in density would continue for the next 10 years.

In 1975, Moore spoke at an IEEE International Electron Devices Meeting, where he looked ahead at the next 10 years and considered how the slow down in the current density gains changed his prediction. The new prediction indicated a doubling every two years. While Moore did not have concrete data in his prediction, it has remained mostly accurate since 1975. Even though not a true law, like the laws of physics, his prediction become to be known as "law" within the electronics industry. Moore's prediction has been considered in guiding planning and research and development. There is some thought in the industry that his prediction has kept the goal post moving, bringing new ideas and technology to bear to keep up with Moore's Law. Some wonder if the industry kept technological growth down due to Moore's Law, and whether the prediction became a virtual ceiling to development (**figure 1.36**).

Figure 1.36—Moore's Law: Transistors per Microprocessor

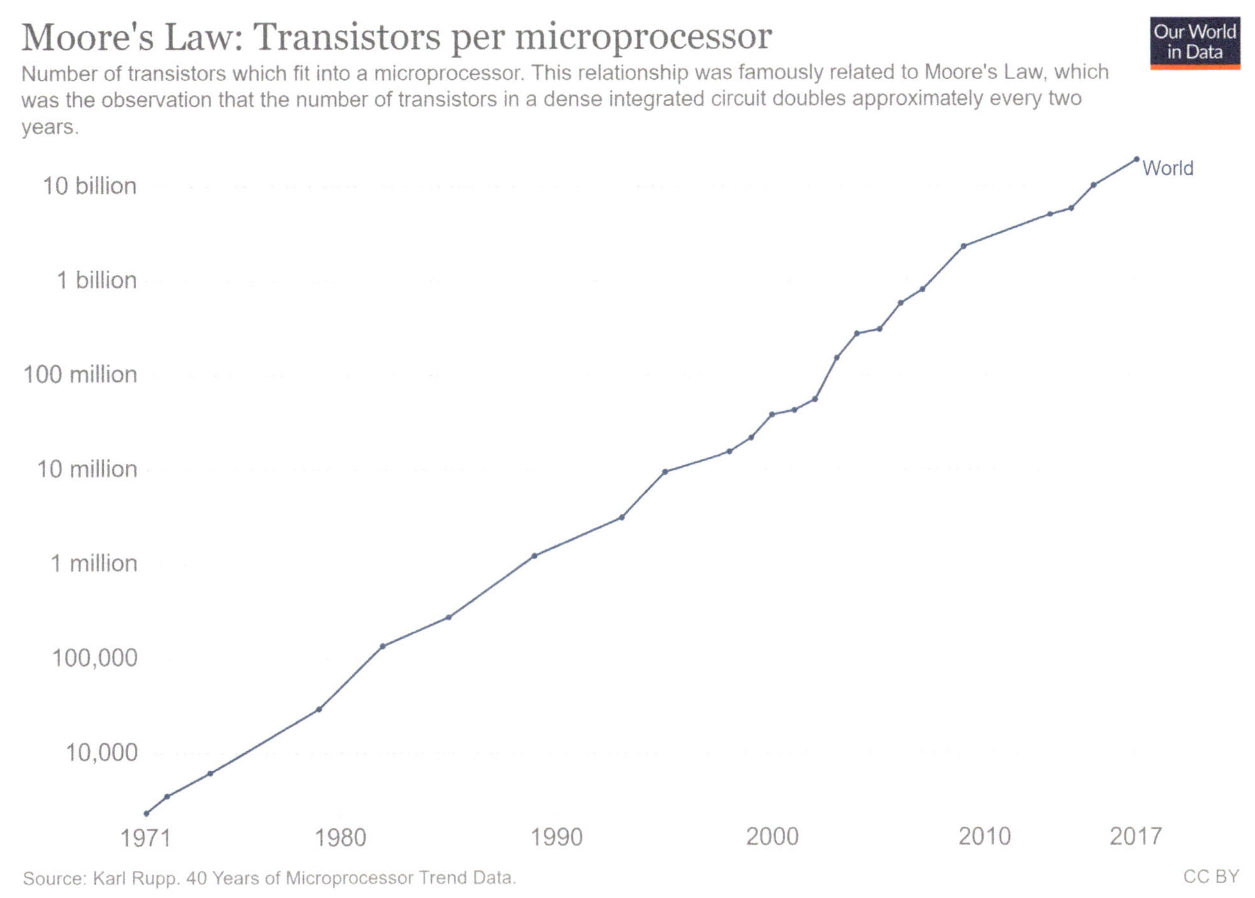

Source: Our World in Data, "Moore's Law: Transistors per Microprocessor," https://ourworldindata.org/grapher/transistors-per-microprocessor. Text available under the Creative Commons International 4.0 license.

Processors

A central processing unit (CPU) is a solid-state device based on what it is made of, and an IC because it is made with multiple electronic devices on the same piece of material. It is also a microprocessor which can use registers, clocking, digital circuits, binary inputs and calculations to enable programs to take and give data from a computer.

For general-purpose use, the CPU for most computers is designed to run programs, and to do it fast. For a personal computer meant just for email and basic web browsing, a general CPU is more than adequate. However, for uses requiring a workstation-type computer for developing CAD, 3D models, or editing video, a general CPU might not be sufficient. For these uses, special processing units have been created:

- A digital signal processor (DSP) is specific to audio and telecommunication needs.
- Graphics processing units (GPUs) are made to render high-quality images and video files.
- Microcontrollers are very useful for embedded systems, such as in fleet vehicles and process control applications.

CPU architecture is probably more known in the context of video game consoles. Early consoles operated with an 8-bit or 16-bit architecture. This nomenclature had to do with the CPU and the associated subsystems within it.

An 8-bit CPU and ALU architecture includes components based on registers, address buses, and data buses of that size. Many microprocessor architectures have had one or more of the subsystems with an ability to have more or less readable bits. For example, many of the early architectures were 8-bit systems, but the memory addresses were 16 or 32 bits; or the software might have been written for 32 math operations even though it was processed 8 bits at a time.

8-bit architectures were just the beginning. **Figure 1.37** describes many different chip designs.

A 12-bit CPU and ALU architecture, means those components are based on registers, address buses, or data buses of that size. A 16-bit CPU and ALU architecture, means those components are based on registers, address buses, or data buses of that size. A 32-bit CPU and ALU architecture, means those components are based on registers, address buses, or data buses of that size. A 64-bit CPU and ALU architecture, means those components are based on registers, address buses, or data buses of that size.

Figure 1.37—Moore's Law: The Number of Transistors on Integration Circuit Chips

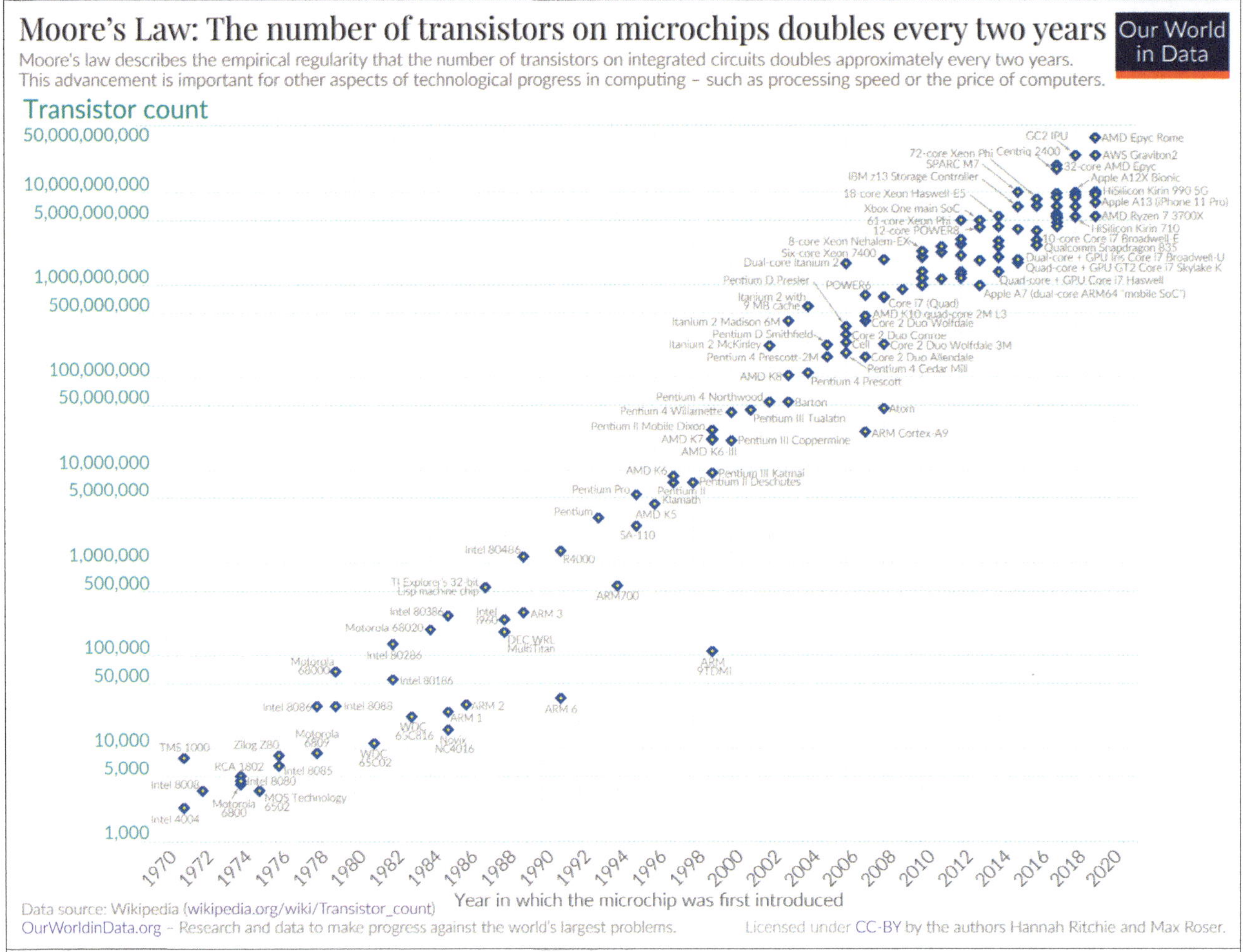

Source: Our World in Data, "Moore's Law: The number of transistors on integration circuit chips," https://ourworldindata.org/technological-progress. Text available under the Creative Commons International 4.0 license.

The first 32-bit system, meaning 32-bit data paths, 32-bit buses and 32-bit addresses, was the AT&T-Bell Labs BELLMAC-32A. The story is interesting because in 1979 and 1980, when the chip was developed, there was no CAD design software like there is today. The development team created a 20X20-foot drawing of the chip diagram in the building's entrance lobby. They literally got down on their hands and knees with colored pencils to make sure that none of the traces were overlapping and they were routed where they needed to be.[3]

The number of transistors and the architecture of single chips has not yet reached its maximum limits. However, getting to those higher numbers and smaller architectures has become increasingly expensive over the years. Engineers realized that the efficiency of existing chips could be increased by using what is already installed in a better way. This gave way to additional technologies such as hyperthreading and multi-core processors.

Hyperthreading (HT) was developed by Intel, and with it, a single CPU core can work on two problems at the same time. That does not mean twice as much work is being done, but rather all its capacity is being used by working on multiple simpler problems at once.

Multi-core processors work much the same way as simultaneous multithreading, but instead of just increasing the efficiency of a single core, multiple cores (CPUs) are placed onto the same chip structure. This allows each core to use hyperthreading and to work simultaneously on different tasks.

1.4.2 Memory

Random access memory (RAM) is a type of memory that can read and write data very quickly. This is the type of memory used for main memory, which can be directly accessed by the CPU. Machine and program data that is currently being worked on is stored in this memory. Today, almost all types of RAM are made of ICs, specifically metal oxide semiconductor chips that contain memory cells. The memory cell is a core component in memory devices. It is simply an electronic circuit, which can store binary information, 1-bit per circuit or cell. It can store a 1 with high voltage and store a 0 with low voltage. The value can be retrieved by reading the cell, and is maintained as long as there is power, until it is changed by the set/reset process. There are a couple of types of RAM.

Static RAM (SRAM) where the memory cell is a flip-flop circuit comprised of two stable states giving it the ability to store binary information. The good thing about SRAM is low power use, but the bad thing is that it has a low storage density and is expensive.

Dynamic RAM (DRAM) uses the charging and discharging of capacitors to store 1s and 0s. Capacitors inherently drain power over time, which means that the memory cells have to recharge. Almost exactly opposite of SRAM, the downside of DRAM is that it uses more power but has higher storage density and costs less.

An additional improvement was made to DRAM which resulted in synchronous dynamic random-access memory (SDRAM). The improvement changed the operation of its external pin interface, controlled by an externally-supplied clock signal, which controls the incoming commands to the memory. The commands can be "pipelined," which is the ability for memory to receive an additional command before the previous command is done processing. The memory is also broken up into banks, and with separate banks of memory, separate commands can operate on each bank at the same time. This allows for higher concurrent data reads and writes, which equate to higher transfer rates.

[3] Allen, G.; "On The BELLMAC-32, And Perhaps the World's Largest Plotter Pen Drawing", https://greg.org/archive/2011/12/09/on-the-bellmac-32-and-perhaps-the-worlds-largest-plotter-pen-drawing.html

Figure 1.38—Double Data Rate SDRAM (DDR SDRAM) Memory

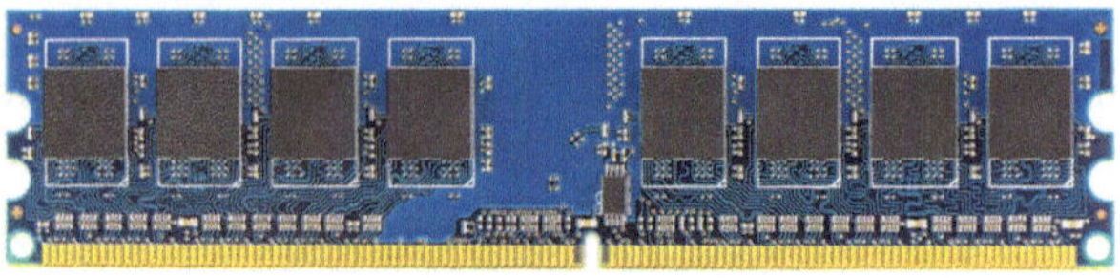

Source: Mkos83; Getty Images

Advancements for SDRAM included double data rate SDRAM (DDR SDRAM), illustrated in **figure 1.38**. This allowed memory to be read on the up and down cycle of the clocking signal. DDR kept growing, not in actual speed, but in the amount of data that could be retrieved in a single cycle. This advancement continued with DDR2, DDR3, DDR4 and DDR5.

Another use for SDRAM is in the synchronous graphics RAM SGRAM device. This type of RAM was specially designed for graphics cards and completing tasks such as texture memory and framebuffers.

Read-only memory (ROM) is a type of non-volatile memory, which does not lose what is stored, if power is removed. More than just removing power, data stored in ROM cannot be changed after manufacturing the device.

ROM is memory that is hard-wired, such as a diode matrix or a mask ROM integrated circuit. A discrete circuit could be changed by completely replacing the chip or component. There are also bodge wires that could be added, but neither of these options is easy. Bodge are soldered to a PCB to fix or change a trace, if something was broken or designed incorrectly. Any errors or software changes must have a whole new device as a replacement.

Other chips, called erasable programmable read-only memory (EPROM) and electrically erasable programmable read-only memory (EEPROM), can be erased and re-programmed. That usually requires a special piece of equipment made to reprogram the chips, which can be expensive and hard to get. Erasing and re-programming is a very slow process, and can only be done a few times.

1.4.3 Drives

At the bottom of the memory hierarchy resides cheaper, larger volume storage devices. A hard disk drive (HDD) is an electro-mechanical data storage device. Electro-mechanical is partially electrical and partially mechanical. The electronic part uses magnetic storage to store and retrieve data. The mechanical part uses hard, fast spinning "platters" coated with magnetic material. The platters look like the foil side of a compact disk (CD).

Data can be read and written in any order, and this is done with magnetic read/write heads attached to a very fast-moving actuator arm. HDDs are a type of non-volatile storage, retaining stored data even when powered off.

The primary characteristics of an HDD are its capacity and performance. Capacity can be affected by the file system and the operating system (OS). HDD makers usually state their capacities in decimal terabytes (powers of 10), but most operating systems state the capacity in binary terabytes.

Performance can be defined with a few factors:

- Average access time: The time required to move the heads to a track or cylinder.
- Average latency: The time required for the desired sector to move under the read head and is a function of the physical rotational speed in revolutions per minute.
- Data rate: The speed at which the data is transmitted.

There are two common sizes or form factors for HDDs: 3.5-inch, for desktop computers, and 2.5-inch, primarily for laptops. HDDs are connected with interface cables such as parallel ATA (PATA), serial ATA (SATA), USB or serial attached SCSI (SAS).

Solid state drives (SSD) are increasingly common. Presently, SSDs still cost more than their alternatives, but prices continue to decrease since release. In applications where storage density (space) is important, such as in laptops, tablets and phones, cost is less of a deciding factor. Additionally, the durability is much greater with an SSD as there are no moving parts. In small, battery-powered devices, the fact that SSDs use much less power is also very attractive. Other forms of storage such as SD cards and non-volatile memory express (NVMe) controller cards are becoming more prevalent for server implementations versus spinning disks. NVMe or non-volatile memory host controller interface specification, is an open standard for accessing non-volatile storage media attached via the PCI Express bus.

1.4.4 Motherboard, Ports and Peripherals

A motherboard is also sometimes known as the main circuit board or system board. The "main" PCB of a computer or system is often called the mainboard instead of motherboard. Sometimes that works, but generally, the difference is this: a motherboard is a PCB with expansion capabilities. A mainboard is generally a PCB which does not expand and has all of its functionality directly attached to its "mainboard" (**figure 1.39**).

Figure 1.39—Mainboard

Source: Chi Rawu Th Lek Kracang/EyeEm; Getty Images

Most computer motherboards have a significant number of the main sub-systems connected directly to them. These sub-systems are typically the CPU, the chipset's input/output and memory controllers and interface connectors. Some of the interface connectors and peripherals can be connected as separate interface cards, also called daughtercards. Here is a short list of some examples of those types of cards:

- Hard drives, solid state drives and other storage

- Video cards
- Network cards
- Sound cards
- Cards providing other expansion ports, USB or FireWire slots

Figure 1.40 shows a common microATX motherboard.

Figure 1.40—microATX Motherboard

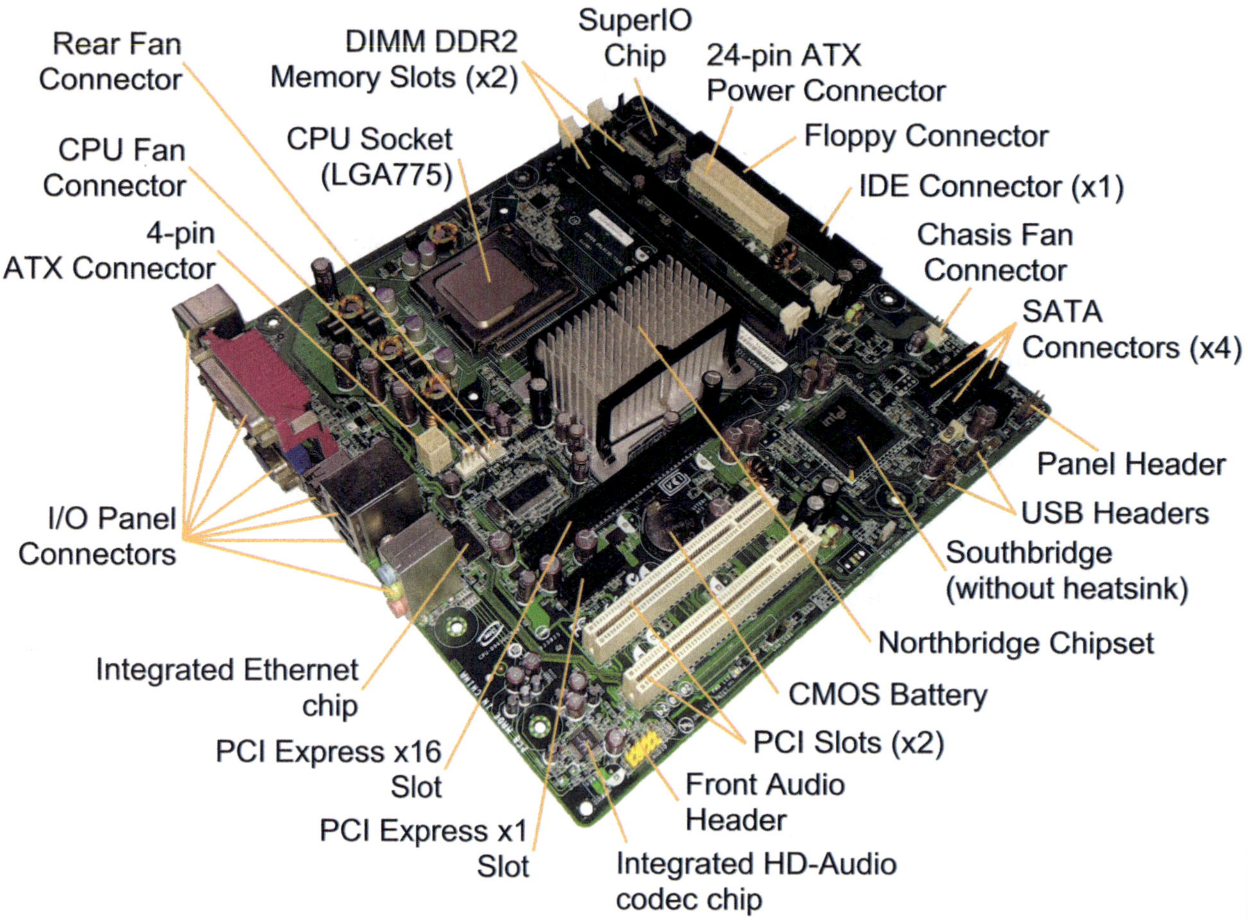

Source: Wikimedia Commons, 21 November 2019, https://commons.wikimedia.org/w/index.php?curid=84225711. The figure is available under the Creative Commons Attribution-Share Alike 3.0 Unported license.

Most motherboards now include a host of interfaces, connectors, expansion and additional processing chips. Some additional items include sockets (or slots) in which one or more CPUs can be installed, and memory slots for the main memory—most common today are DIMM modules containing dynamic random access memory (DRAM) chips. There are also chipsets, which are comprised of a pair of chips—the Northbridge and the Southbridge. As seen in **figure 1.41**, they make up the connections between the CPU's front-side bus, main memory and peripheral buses.

Figure 1.41—Chipset

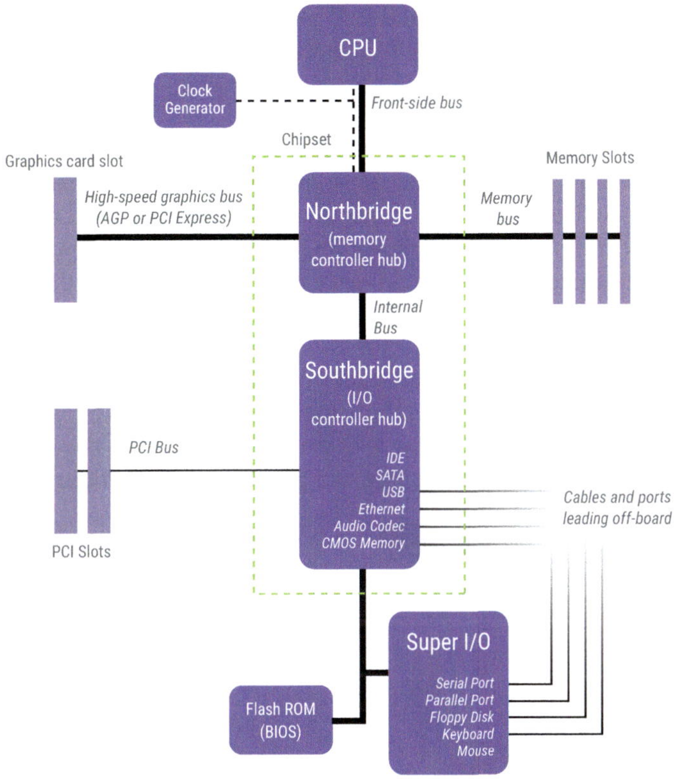

Source: Wikimedia Commons, "Motherboard Diagram," https://commons.wikimedia.org/wiki/File:Motherboard_diagram.svg, 11 April 2007. The figure is available under the Creative Commons Attribution ShareAlike license."

Non-volatile memory chips, typically flash-read only memory (ROM), contain firmware or basic input-output system (BIOS). Most designs now use BIOS, stored in a type of ROM chip called an electrically-erasable programmable read-only memory (EEPROM) chip either soldered or in a socket on the motherboard, which boots the operating system (OS).

There are slots for expansion cards, such as the PCI and advanced graphics port (AGP) slots. The AGP or PCI express slot allows for adding high throughput video graphics processing cards. The PCI slots allow for adding a wide array of add on cards and ports, such as sound, video and storage options.

Connections for storage drives are now mostly SATA, but may also be IDE, SCSI or SAS.

Power Supplies

A power supply unit (PSU) does a straightforward job which is vitally important; without a stable and accurate PSU, a computer will not function. The PSU converts AC power into low-voltage, regulated DC power. Most modern computers use a switched-mode power supply. Some PSUs use manual switches, like little dip switches, for selecting the input voltage.

A switched-mode power supply (SMPS) is a PSU that uses a switching regulator to convert electrical power efficiently. The previous type of PSU was a linear power supply, which used a heavy step-down transformer. SMPSs are much smaller and lighter than a linear supply, mostly because the transformer is smaller in size and weight. These

PSUs were very inefficient as much of the energy expended by them was wasted in heat and conversion loss; it was always on and "burning" power, even if there is little load. The pass transistor of an SMPS switches on and off constantly (thousands of times per second). This allows the SMPS to spend less time and energy in high dissipation (heat) transitions, thus reducing the wasted energy. There is no perfect SMPS that dissipates no power, but there are hypothetical designs to achieve one.

Most computers and electronic devices require one or more stable DC voltages. To achieve and maintain a quality voltage, a regulation circuit is used to vary the on/off time. Most modern personal computer PSUs follow the ATX (Advanced Technology eXtended) specification. The ATX specification was created and put out by Intel Corporation. The ATX specification was to resolve the issue of multiple varying form factors and no way to create sufficient standardization, especially since Intel wanted systems to properly run and operate their CPUs.

The specification controls gives best practices for form factor (size and shape) and voltage tolerances. As long as an ATX power supply is connected to power, it gives a 5 Volt standby (5VSB) or control voltage. This allows the computer and certain peripherals to keep standby or sleep functions powered. In some cases, that might mean the universal serial bus (USB) ports are active and have power. For example, some USB devices require standby or control voltage to keep running to charge a cell phone. Below is a picture of a PSU (**figure 1.42**).

Figure 1.42—PSU

Source: Alarich; Getty Images

ATX-specified PSUs get a signal from the motherboard to turn on and off. The PSU, when first connected to power, will wait and provide a signal to the motherboard, which tells the computer when the DC voltages are stable. This is how the computer and connected devices know when it is safe to boot. Intel's design guide, shows multiple connectors and pins for providing the power and control.[4]

In 2020, Intel announced a new ATX standard called ATX12VO to address increasing governmental standards regarding energy efficiency. The previous standard was version 2.4, released in 2013. The 2.4 version added the extended ATX (EATX) requirements.

Ports and Peripherals

Ports and peripherals are the connectors and interfaces on the motherboard. To make all connectors easier to locate, a color coding scheme was devised. The design specification was developed by Microsoft and Intel who were able to spread it as the "standard." The system for color coding was not maintained after 2001, but almost all computers still follow the same basic color scheme.

[4] Intel, *ATX12VO (12V Only) Desktop Power Supply: Design Guide*, Version 002, May 2020, https://www.intel.com/content/dam/www/public/us/en/documents/guides/single-rail-power-supply-platform-atx12vo-design-guide.pdf

The PC System Design Guide (also known as the PC 97, PC 98, PC 99, or PC 2001 specification) is a series of hardware design requirements and recommendations for IBM PC-compatible personal computers, compiled by Microsoft and Intel Corporation during 1997–2001 (**figure 1.43**). PC System Design Guides were aimed at helping manufacturers provide hardware that made the best use of the capabilities of the Microsoft Windows operating system, and to simplify setup and use of such computers.

Figure 1.43—PC System Design Guide Version and Release Dates

Version	Release Date
PC 97	February 9, 1998
PC 98	December 31, 1998
PC 99	July 14, 1999
PC 2001	November 2, 2000

Source: Wikipedia, PC System Design Guide, https://en.wikipedia.org/wiki/PC_System_Design_Guide#:~:text=The%20PC%20System%20Design%20Guide,Intel%20Corporation%20during%201997%E2%80%932001. Text available under the Creative Commons Attribution ShareAlike 3.0 Unported license.

PC 97 – first version. Introduced color code for PS/2 keyboard (purple) and mouse (green) connectors

PC 98 – for computers with Windows 98 or Windows 2000. Required:

- 200 MHz Pentium processor with MMX technology (or equivalent performance)
- 256 KB L2 cache
- 32 MB RAM (recommended: 64 MB of 66 MHz DRAM)
- ACPI 1.0
- Fast BIOS power-up
- BIOS Y2K compliance
- PXE preboot environment

PC 99 – Required:

- 300 MHz CPU
- 64 MB RAM
- USB
- Comprehensive color-coding scheme for ports and connectors

Strongly discouraged:

- Non plug-and-play hardware
- ISA slots

PC 2001 – Final version. Required IO-APICs. Some “legacy” devices like ISA expansion slots and use of MS-DOS are forbidden.

Figure 1.44 shows the color-coding scheme for connectors and ports.

Figure 1.44—Color-coding Scheme for Connectors and Ports

Color	Function	Connector on PC
User input		
Green	PS/2 mouse / pointing device	6-pin mini-DIN female
Purple	PS/2 keyboard	6-pin mini-DIN female
Gold	Game port / MIDI	15-pin D female
General input/output		
White	USB 1 low speed and full speed	USB Type A receptacle
Black	USB 2.0 high speed	USB Type A receptacle
Sky blue	USB 3.0 SuperSpeed	USB Type A SuperSpeed receptacle
Yellow	USB 3.0 SuperSpeed Always On even if the PC is turned off	USB Type A SuperSpeed receptacle
Red	High-power SUB (more than 500 mA current available)	USB Type A receptacle
Grey	IEEE 1394 (FireWire)	6-pin FireWire 400 or 9-pin FireWire 800
Burgundy	Parallel port	25-pin D female
Teal or turquoise	Serial port	9-pin D male
Video		
Blue	Analog monitor	15-pin VGA female
White	Digital monitor	DVI female
Yellow	S-video	4-pin mini-DIN
Yellow	Composite video	RCA jack
Black	Digital audit/video	HDMI or DisplayPort female
Audio		
Pink	Analog microphone audio input (mono or stereo)	3.5 mm TRS
Light blue	Analog line level audio input	3.5 mm TRS
Lime green	Analog line level audio output, front stereo (speakers or headphones)	3.5 mm TRS
Black	Analog line level audio output, rear stereo (surround speakers)	3.5 mm TRS
Orange	Analog line level audio output, center and subwoofer (surround speakers)	3.5 mm TRS
Silver	Analog line level audio output, side stereo (surround speakers)	3.5 mm TRS
Orange	Digital audio output (coaxial S/PDIF)	RCA connector
Black	Digital audio output (optical S/PDIF)	TOSLINK

Source: Wikipedia, "PC System Design Guide," 28 April 2020, https://en.wikipedia.org/wiki/PC_System_Design_Guide#:~:text=The%20PC%20System%20Design%20Guide,Intel%20Corporation%20during%201997%E2%80%932001. Text available under the Creative Commons Attribution-ShareAlike 3.0 Unported license.

1.5 Software Components of Computing Devices

Hardware generally does very little work without an instruction on when and how to complete that work, and all those instructions come from software. This section introduces many types of software for a computer, starting with the most basic instructions to hardware on how to start up and then test itself. The hardware/computer then needs software to continue running. This is the operating system (OS), with which users can interact and change for their needs. Operating systems seldom have every function that a user might need for their computer, so additional applications can be installed to add functionality and better ability to interact and change things on their specific hardware.

1.5.1 Firmware

Basic input/output system (BIOS) is firmware used to perform the initial boot up of hardware. This is sometimes known as the power-on startup or the power-on self-test (POST). Firmware is lower-level microcode which helps to load the machine instructions. It bridges the gap between hardware and software which lends the name "firmware." The BIOS firmware is the very first thing that loads after power is turned on. The BIOS is installed on a computer's motherboard at the time it is manufactured. The name comes from the basic input/output system used in the CP/M operating system in 1975.

The BIOS, after powering and testing the hardware, then loads the boot loader from secondary memory and signals the OS to load. Back in the day of DOS (disk operating system) the BIOS provided a hardware abstraction layer (HAL) for input/output (I/O) devices which created a standard for those devices to communicate with the OS and applications. Most operating systems today do not use BIOS after loading, the OS can directly communicate with hardware.

The primary functions of the BIOS during POST are:

- Verify CPU registers
- Verify the integrity of the BIOS code itself
- Verify some basic components like DMA, timer and interrupt controller
- Find size and verify the system main memory
- Initialize the BIOS
- Pass control to other specialized extensions
- Identify, organize and select which devices are available for booting
- Discover, initialize and catalog all system buses and devices
- Provide a user interface for the system's configuration
- Construct thc cnvironmcnt for the OS

Original POST code (from IBM) has different beeps for particular errors (**figure 1.45**).

Figure 1.45—Meaning of Beeps

Beeps	Meaning
1 short beep	Normal POST—system is okay
2 short beeps	POST error—error code shown on screen
No beep	Power supply, system board problem, disconnected CPU or disconnected speaker
Continuous beep	Power supply, system board, RAM or keyboard problem
Repeated short beeps	Power supply, system board or keyboard problem
1 long, 1 short beep	System board problem
1 long, 2 short beeps	Display adapter problem (MDA, CGA)
1 long, 3 short beeps	Enhanced Graphics Adapted (EGA)
3 long beeps	3270 keyboard card error

Source: PCHell.com, "Power On Self Test - POST Beep Codes," https://www.pchell.com/hardware/beepcodes.shtml

Because the BIOS directly interfaces with the hardware, it needed to be written specifically for each computer or set of hardware. The BIOS used to be written or made into a ROM chip on the motherboard. On most computers today, the BIOS file(s) are written to flash memory, which means it can be updated without changing a chip.

1.5.2 Operating System Architecture

An operating system (OS) is the interface between the computer's hardware and a user. More specifically, a user's ability to run and interact with a program. The main parts of an OS are the:

- Kernel
- Shell
- Applications

It may help to think about these parts and layers as to whether they are closer to or further from the hardware of the system. These OS parts are shown in **figure 1.46**.

Figure 1.46—Operating System Parts

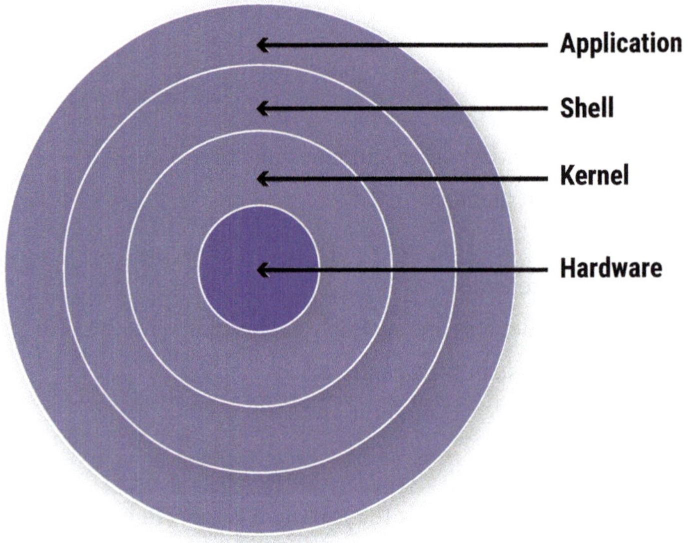

Of the three main parts of the OS, the kernel is the primary piece. The kernel controls and manages all system resources, and talks back and forth with the hardware and the software. It serves as the bridge and the lowest abstraction layer for the hardware. For a program (application) to run, it often needs to control how the hardware will perform. It is not a good practice to allow random software to have direct control of the hardware. Instead, the kernel makes the hardware available to the applications through inter-process communication (IPC) and system calls.

The shell, which is called this because it is the outermost layer for accessing the OS, is the user interface to the computer and OS. Some OSs only have what is known as a command line interface (CLI), which is a text-based interface or editor. Some OSs have a graphical user interface (GUI) as their primary interface. GUIs have visual, point-and-click, interfaces. GUI interfaces can have certain limitations, such as not having all the commands or options to be run. It is easy to script repetitive tasks to be performed in the CLI that are not possible in a GUI. For this reason, most OSs, even if primarily GUI-based, have a CLI shell, as well.

OSs need a place to store all data that is added and changed. For example, when an application is installed. There is data about where it is stored, how it runs and how to start it. Then, when a user runs it, there can be personal and global settings and preferences for the application. Most OSs use a type of database to store and locate this data.

Databases, like a single spreadsheet, are tables of sorts. They have rows and columns where data can be stored. The data can be easily found using the column letter and the row number. But for different data, there might be a need for multiple spreadsheets, and pieces of data from each of the sheets related back to one main item. An example of a simple database is multiple spreadsheets linking data back and forth between the spreadsheets. Keeping track of all the sheets and which data is stored in each one requires database languages and programs that help make the process of storing, finding and organizing the data much easier.

As an example of an OS database, Microsoft Windows uses a database to store all of its data. The Windows database is called the registry. The Windows registry is a hierarchical database that stores low-level settings for the Microsoft Windows OS and applications that opt to use the registry. The kernel, device drivers, services, Security Accounts Manager and user interface can all use the registry. In other words, the Windows registry contains information, settings, options and other values for programs and hardware installed on all versions of Microsoft Windows operating systems.

Following are some of the most widely used types of OSs. These are not products or marketing names of OSs, but rather indicate operational types or classifications of OSs:

- Simple batch system
- Multiprogramming batch system
- Multiprocessor system
- Desktop system
- Distributed operating system
- Clustered system
- Realtime operating system
- Handheld system

Simple Batch Systems

In a simple batch system, there is no direct interaction between the user and the computer. The user has to submit a job, written on cards or tape, to a computer operator.

Multiprogramming Batch Systems

In multiprogramming batch system, the OS picks up and begins to execute one of the jobs from memory. When this job needs an I/O operation, the OS switches to another job since the CPU and OS are always busy. Jobs in the memory are always less than the number of jobs on the disk, known as the job pool.

Desktop Systems

In the past, CPUs and PCs lacked the features needed to protect an operating system from user programs. PC operating systems therefore were neither multiuser nor multitasking. However, the goals of these operating systems have changed over time; instead of maximizing CPU and peripheral use, the systems opt to maximize user convenience and responsiveness. These systems are called desktop systems and include PCs running Microsoft Windows and the Apple Macintosh. Operating systems for these computers have benefited in several ways from the development of operating systems for mainframes.

Distributed Operating System

The motivation behind developing distributed operating systems is the availability of powerful and inexpensive microprocessors and advances in communication technology.

These technology advancements have made it possible to design and develop distributed systems comprised of many computers that are interconnected by communication networks. The main benefit of a distributed system is its low price/performance ratio.

Following are the two types of distributed operating systems:

- Client-server systems
- Peer-to-peer systems

Clustered Systems

Like parallel systems, clustered systems gather together multiple CPUs to accomplish computational work. Clustered systems differ from parallel systems, however, in that they are composed of two or more individual systems coupled together.

The definition of the term clustered is not concrete; the generally-accepted definition is that clustered computers share storage and are closely linked via local area network (LAN) networking.

Real Time Operating System

A real-time operating system is defined as an operating system known to give maximum time for each of the critical operations that it performs, like OS calls and interrupt handling.

Handheld Systems

Handheld systems include phones, tablets and readers with connectivity to a network, such as the Internet. They are usually smaller in size, so handheld devices typically have less memory, include slow processors and feature small display screens.

1.5.3 Operating System Types

This section describes the types of OSs that are used in computer systems today.

Microsoft

Microsoft® develops entire families of Windows® operating systems that are specific to types of companies and uses.

The following are Windows families being actively developed:

- Windows: The operating system for mainstream personal computers, tablets and smartphones. The latest version is Windows 10.
- Windows Server: The operating system for server computers. The latest version is Windows Server 2019.
- Windows PE: A lightweight version, used for installing Windows on bare-metal computers, recovery or troubleshooting purposes. The latest version is Windows PE 10.
- Windows IoT (previously Windows Embedded): Microsoft developed Windows CE for a device that was small and only had enough resources for specific functions—not intended to be a full working desktop computer. To understand the different names of this software requires a bit of history. Windows CE, as it was originally called, was renamed Windows Embedded Compact and was folded under the Windows Compact trademark. This trademark also includes other software packages that Microsoft uses for similar purposes. Those software packages consist of Windows Embedded Industry, Windows Embedded Professional, Windows Embedded Standard, Windows Embedded Handheld and Windows Embedded Automotive.

Apple (Mac)

Apple® makes macOS®, which has been called both Mac OS X and later OS X. The current name is macOS, and is a family of proprietary graphical operating systems developed and marketed by Apple since 2001. It is the primary operating system for Apple's Mac® computers.

macOS is the second major series of Macintosh operating systems. The first is often referred to as classic Mac OS for reference purposes. The final release was Mac OS 9, which rolled into the desktop version, Mac OS X 10.0. For years, Apple has had unique names for their OS releases, and in some cases funny stories about how the names came to be. First, the releases were named after big cats, then the releases started being named after locations in California:

- Mac OS X 10.0 Cheetah
- Mac OS X 10.1 Puma
- Mac OS X 10.2 Jaguar
- Mac OS X 10.3 Panther
- Mac OS X 10.4 Tiger
- Mac OS X 10.5 Leopard
- Mac OS X 10.6 Snow Leopard
- Mac OS X 10.7 Lion
- OS X 10.8 Mountain Lion
- OS X 10.9 Mavericks
- OS X 10.10 Yosemite
- OS X 10.11 El Capitan
- macOS 10.12 Sierra
- macOS 10.13 High Sierra
- macOS 10.14 Mojave
- macOS 10.15 Catalina
- macOS 11.0 Big Sur

Apple changed the OS naming to "macOS," which brought synergy to the naming of their other operating systems, iOS, watchOS and tvOS.

Linux

Linux® is a family of open-source OSs based on the Linux kernel, an OS kernel first released by Linus Torvalds on September 17, 1991. Distributions include the Linux kernel, supporting system software and libraries. A distribution is a package of software that makes up the entire OS. Anyone can create a whole new distribution by creating a new GUI or changing the OS's supported hardware. The Linux kernel is open-source, but many of the associated libraries and software packages for drivers and GUI might not be, or may fall under something called the GNU Project.

The GNU Project (stands for GNUs, not Unix) was started in the early 1980s as a rebellion against the creators of Unix, with the goal of making a completely free OS. This did not happen until Torvalds created the Linux kernel. Many distributions use the word "Linux" in their name, but the Free Software Foundation (which is the legal non-profit company that was established to promote and help the GNU Project) prefers the name GNU/Linux to impart the importance of GNU software. This has caused some controversy.

Some of the more well-known Linux distributions are:

- Debian
- Fedora

- Ubuntu
- Kali

Some distributions that have a paid aspect, even if just for support:

- Red Hat Enterprise Linux
- SUSE Linux Enterprise Server

Many Linux distributions have a GUI or windowing system, such as X11 or Wayland, and a desktop environment such as GNOME or KDE Plasma. Distributions intended for servers may omit graphics altogether, or include a solution stack such as LAMP.

Linux was originally developed for personal computers based on the Intel x86 architecture, but has since been ported to more platforms than any other operating system. Linux is:

- The leading operating system on servers
- The leading operating system on big iron systems such as mainframe computers
- The only OS used on TOP500

Linux also runs on embedded systems, such as:

- Routers
- Automation controls
- Smart home technology
- Televisions
- Automobiles
- Digital video recorders
- Video game consoles
- Smartwatches
- Smartphones and tablets

1.5.4 Operating System Configuration

Considering all the devices and OS architectures that exist, software must be written so all things function correctly. With so many varieties of hardware and operating systems, drivers are the software written to interface between the hardware and operating systems. A driver understands the specific commands that must be given to a device to produce whatever work the device does. The driver must also understand the commands and data that will come from the OS and applications. This translation of languages must happen for all devices, such as:

- Printers
- Video adapters
- Network cards
- Sound cards
- Computer storage devices (ATA, SATA, SCSI)
- Image scanners
- Digital cameras

The translation is often called abstraction. Microsoft even denotes an entire layer of OS development to this called the hardware abstraction layer (HAL). The HAL allows an OS to interact with hardware devices.

Common levels of abstraction for device drivers include:

For hardware:

- Interfacing directly
- Writing to or reading from a device control register
- Using some higher-level interface
- Using another lower-level device driver
- Simulating work with hardware

For software:

- Allowing the operating system direct access to hardware resources
- Implementing only primitives
- Implementing an interface for non-driver software
- Implementing a language

So choosing and installing the correct device drivers for given hardware is often a key component of computer system configuration.

Utilities and other applications have the same items that need to be written into them. An application, for instance, must have routines. The logic in software commands that is used over and over for an action to happen is called a routine, for example, to move data from secondary storage to main memory, and to tell the CPU in what way to process the data it needs. All these items make changes or add to the OS functionality to allow additional software to run and use the system resources.

1.5.5 Application Software

Application software is defined as applications written for a computer for end users to use and produce work. Think of any software a person might use on the computer, and it is almost certainly an application. Even if it is built-in to the OS, it is still a written application that is bundled and runs on the computer just the same. Any app store lists the categories available.

Applications have other effects on the market today. Many applications drive users to purchase specific systems or hardware, and some consider it being locked into a specific ecosystem. This happens so frequently that a term has been dubbed for this type of application: killer application or killer app.

Application and Software Licensing

Application software is typically broken down into 2 main classifications:

- Closed-source vs open-source software applications
- Free or proprietary software applications

Closed-Source and Open-Source Software Applications

Closed-source software is where intellectual property (e.g., computer source code) is not made available to the general public by its creators.[5] Generally speaking, the comparison of free vs. proprietary software is analogous to closed vs. open source. However, the same is not true for free software and open-source as they convey different philosophies.

Open-source software means all software that is distributed as "free software," "open-source software" or under a "copyleft" agreement or a similar licensing or distribution model. It also includes software subject to the terms of any license which requires, as a condition on the use, copying, modification, and/or distribution of such software that such item, or other software incorporated into, derived from, or distributed with such item, (i) be disclosed or distributed in source code form, (ii) be licensed for the purpose of making derivative works, or (iii) be redistributed at no or minimal charge.[6] "Open Source Software" includes, without limitation, software distributed under the GNU General Public License, GNU Lesser General Public License, Mozilla Public License, BSD licenses, the Netscape Public License, the Sun Community Source License, the Sun Industry Standards License and the Apache License.

Free or Proprietary Software Applications

Richard Stallman started the GNU Project and the Free Software Foundation and is a staunch proponent that free equals freedom. Supporters of free or libre software advocate that the software is free and any particular user has the freedom to do whatever they want to include use, modification, redistribution of copies (even for a fee) and redistribution of modified versions.[7] Conversely, open-source software refers to the licensing, and many end-user licensing agreements stipulate what can and cannot be done with the software.[8] In this regard, it is been described that free software is equivalent to free speech, whereas open-source software is like a free beverage.[9]

1.6 Data and File Systems

Similar to an OS database and memory addressing, all data that is stored must be able to be found and used again. Every type of storage is going to have data read and written to it, and will have some kind of file system dictating how and where the data will be stored in that particular type of memory. If there was no control, the entire storage system would be a long chain of data; no start, no stop and no way of finding the data.

Before computers, when data processing was done by hand and on paper, each group of data was called a file. Often the paper-based data was stored in a folder, which is a term still used in computer data processing. In a file cabinet full of "files" and "folders," there had to be an organization method to find them. With digital data, that method of organizing the data is called the "file system."

There are many different types and standards of file systems. Each one has different structure, logic, speed, flexibility, security, size and more. Some file systems have been designed to be used for specific applications.

1.6.1 File System Structure

There are file system types for every kind of storage medium. Just a few are:

- Optical disks
- Disks
- Flash

[5] *Collins English Dictionary: Complete and Unabridged*, "closed source," 12th Edition, 2014, https://www.thefreedictionary.com/closed+source
[6] Law Insider, "Definition of Open Source Software," https://www.lawinsider.com/dictionary/open-source-software#:~:text=Open%20Source%20Software%20means%20(i,redistributed%20only%20free%20of%20enforceable
[7] Stallman, R.; The GNU Project, https://www.gnu.org/gnu/thegnuproject.en.html
[8] The Open Source Initiative, "The Open Source Definition," https://opensource.org/osd
[9] Stallman, R.; "Why Open Source misses the point of Free Software," GNU Operating System, https://www.gnu.org/philosophy/open-source-misses-the-point.html

- Tape
- Database
- Network

The most common file systems are for HDDs, and in many cases are created by the makers of a particular OS.

A disk file system takes advantage of disk storage media's ability to randomly address data in a short amount of time. Additional considerations include the speed of accessing data following that initial request and the anticipation that the following data may also be requested. This permits multiple users (or processes) access to various data on the disk without regard to the sequential location of the data. Some examples are:

- FAT (FAT12, FAT16, FAT32), exFAT
- NTFS
- HFS and HFS+, HPFS
- APFS
- UFS
- ext2, ext3, ext4
- XFS
- Btrfs

Journaling File Systems

Many file system standards use journaling or versioning file systems to write, track and ensure the integrity of data.

A journaling file system keeps a journal of the changes made to the file system. The journal contains a change log, sometimes with the actual data and sometimes with metadata. If there is a system crash, the file system can be preserved or corrected with the notes that are in the journal preventing permanent loss of data.

A versioning file system keeps multiple copies or versions of all files. This also works as a type of revision control, which could allow an older version of a file to be recovered. This type of file system could be resource intensive if the number of copies, changes per amount of time and trivial changes are not throttled.

1.6.2 File Management and Permissions

Most file systems have a way to give or remove permissions or access rights to files. This can be done for specific users and groups of users. Along with who can access something, there are typically also controls for how, or at what level, the access is given or removed. The parameters for this access are generally notated with the terms view, change, navigate and execute. These permissions can be set on an individual file, folder (directory), partition or drive.

Two types of permissions are very widely used:

- Traditional Unix permissions
- Access control lists (ACLs)

Unix Systems Permissions

Permissions on Unix systems are managed in three distinct scopes or classes. These scopes are known as user, group, and others.

Unix systems implement three specific permissions that apply to each class:

- Read permission grants the ability to read a file. When set for a directory, this permission grants the ability to read the names of files in the directory, but not to find out any further information.
- Write permission grants the ability to modify a file. When set for a directory, this permission grants the ability to modify entries in the directory, which includes creating files, deleting files and renaming files. Note that this requires that execute is also set; without it, the write permission is meaningless for directories.
- Execute permission grants the ability to execute a file. This permission must be set for executable programs, to allow the operating system to run them. When set for a directory, the execute permission is interpreted as the search permission: it grants the ability to access file contents and meta-information if its name is known, but not list files inside the directory, unless read permission is set also.

Setting permissions on directories in Unix systems is often misunderstood and frequently done incorrectly.

Access-Control List Permission System

An access-control list (ACL) permission system is a list for each object including who may interact with an object, and what can be done with the object. A listing in an ACL specifies a subject and an operation. For example, a file object has an ACL that contains **Pat: read,write**; **Nira: read**. This gives Pat permission to read and write the file, but gives Nira only read permission.

1.6.3 File Types

A file type, or file format, is the defined way for a specific file to encode data within that container. Container is another word for file, which is the total collection of all the data in the file. Pieces of data are organized to be used and stored by the file format. For example, a portable network graphics (PNG) file, which is a raster image file, is stored in a way that the image will not lose quality. PNG is similar and was developed to be a non-patented version of the graphics interchange format (GIF) file format.

Just like GIF files, many file types are proprietary and/or patented in use. Many are also open-use file types. There are file types for files that are executables (files that start a program), audio files, image files, video files or email files. Here are some of the most common categories and file types of each:

Audio file types (formats):

- **mp3**—MP3 audio file
- **wav**—WAV file
- **wma**—WMA audio file

Compressed file extensions:

- **.7z**—7-Zip compressed file
- **pkg**—Package file
- **rar**—RAR file
- **zip**—Zip compressed file

Disc and media file extensions:

- **bin**—Binary disc image
- **dmg**—macOS X disk image
- **iso**—ISO disc image
- **vcd**—Virtual CD

Data and database file extensions:

- **csv**—Comma separated value file
- **txt**—text file format
- **sql**—SQL database file
- **tar**—Linux / Unix tarball file archive
- **xml**—XML file

E-mail file extensions:

- **email**—Outlook Express email message file
- **eml**—E-mail message file from multiple e-mail clients, including Gmail
- **emlx**—Apple Mail e-mail file

Executable file extensions:

- **bat**—Batch file
- **bin**—Binary file
- **exe**—Executable file
- **jar**—Java Archive file
- **msi**—Windows installer package

1.6.4 Data Types

A data type is an aspect of data that tells the compiler or interpreter (software) how the programmer of the software is going to use the data. Most programming languages support basic data types as illustrated in **figure 1.47**.

Figure 1.47—Data Types and Uses

Data Type	Explanation of How the Software Uses the Data
Integer numbers	Unlike mathematical integers, typical datum in a computer has minimum and maximum possible values.
Floating-point numbers	Refers to the fact that a number's radix point (decimal point, or binary point) can "float:" that is, it can be placed anywhere relative to the significant digits of the number. This position is indicated as the exponent component, and thus the floating-point representation can be thought of as a kind of scientific notation.
Characters	A character is a unit of information that roughly corresponds to a grapheme, grapheme-like unit, or symbol, such as in an alphabet or syllabary in the written form of a natural language.
Boolean	The Boolean data type has one of two possible values (usually denoted true or false) which are intended to represent the two truth values of logic and Boolean algebra
Strings	A string is traditionally a sequence of characters, either as a literal constant or as a kind of variable.

1.6.5 Data Life Cycle

All types of collected or created data has a use or value. After a period of time, it is possible that data can become a liability, using resources and costing money, and at a point it is no longer useful. The data life cycle as shown in

figure 1.48 shows the process of data from the time it is made until it is destroyed or deleted. **Figure 1.49** further explains some of the data life cycle stages.

Figure 1.48—The Information Life Cycle

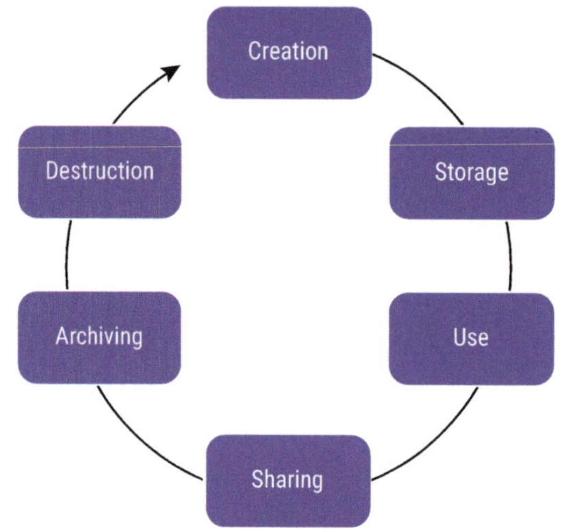

Figure 1.49—Explanation of Data Life Cycle Stages

Data Life Cycle Stages	Explanation
Collect or create	This is when data is made or comes under control. There is cost for every bit of data that is kept. Even if only the space on a computer, that storage has a cost. If the data is important, it should be backed up or synchronized to a cloud location, which has additional costs. Not to mention using and transporting that data.
Secure and store	How important the data is may still come down to cost, but not just the cost in storing it. If the data is important to a business and would negatively affect the business if it was lost, securing the data and marking it as important is critical.
Use	Data is vitally important, but there can be cases where the data becomes a liability. If it is personal, financial or healthcare related, then keeping or using inappropriate data can be costly (in the form of fines, lawsuits, or oversight expenses) for a business.
Disposal	Disposal cannot be just deleting the data off of a computer, it is important to have a plan to move, back up, secure and copy data. It is important to understand where and how all system data is being handled.

1.7 Chapter 1 Knowledge Check

REVIEW QUESTIONS

1. Which of the following types of memory do not lose content when powered off? (Choose all that apply.)
   A. Random access memory
   B. Read only memory
   C. Tertiary memory
   D. Main memory

2. Open-source software and free software are the same. (True or False)
   A. True
   B. False

3. For each of the following, supply the value for the indicated conversion.

428 decimal to binary	
1001 decimal to hex	
11110010010 binary to decimal	
1101 binary to hex	

4. How many bits are in an octet?

5. How many bytes are 48 bits?

Answers on page 68

Chapter 1 ANSWER KEY

Review Questions

1. A. Random access memory
   B. Read only memory. Refer to page 34.
   C. Tertiary memory. Refer to page 33.
   D. Main memory. Refer to page 33.

2. A. True
   B. False. Refer to pages 61-62.

3. For each of the following, supply the value for the indicated conversion. **Refer to pages 19-24.**

428 decimal to binary	110101100
1001 decimal to hex	3E9
11110010010 binary to decimal	1938
1101 binary to hex	D

4. **8. Refer to page 84.**

5. **6. Refer to page 15.**

COMPUTING
FUNDAMENTALS
An ISACA Certificate

Chapter 2:

Networking, Virtualization and Security

Networking, Virtualization and Security

Networking, Virtualization and Security

2.1 Learning Objectives

After completing this chapter, learners will be able to:

1. Identify layers of the Open Systems Interconnection (OSI) Reference Model.
2. Describe network topologies.
3. Explain the functions of OSI model layers.
4. Identify benefits of cloud computing.
5. Identify cloud deployment models.
6. Identify objectives of information security.
7. Describe characteristics of a threat landscape.
8. Describe common types of authentication.
9. Describe functions aligned to security operations.

2.2 Introduction

As part of an even greater expansion, the focus starts moving from within a single system or computer, and out onto the network. What is a network? What is available on a network? How is a network different than the Internet? These questions and many others are covered in this section.

2.3 Networking Concepts

A network is two or more hosts (a computer or similar device) connected by a shared medium to make sharing data, devices, applications and connections to other networks available. **Figure 2.1** shows a simple network.

Figure 2.1—Bus Type Computer Network

2.3.1 Open Systems Interconnection (OSI) Model

The engineers working to create network protocols soon realized that a standard was needed. In the early 1970s, many vendors began creating their own proprietary network protocols. An example was IBM, who throughout the 1970s and 80s, created, published and sold their System Network Architecture (SNA) standard model.

The Department of Defense (DoD), through a 1960s funding project called Advanced Research Projects Agency (ARPA) worked on an experimental networking project connecting and maintaining communication between all military units during war time. The project at that time was called ARPANET and was rolled into other projects, creating an entire DoD model. That R&D became known today as the Transmission Control Protocol/Internet Protocol (TCP/IP) model (**figure 2.2**).

Figure 2.2—OSI Association with the DoD Model

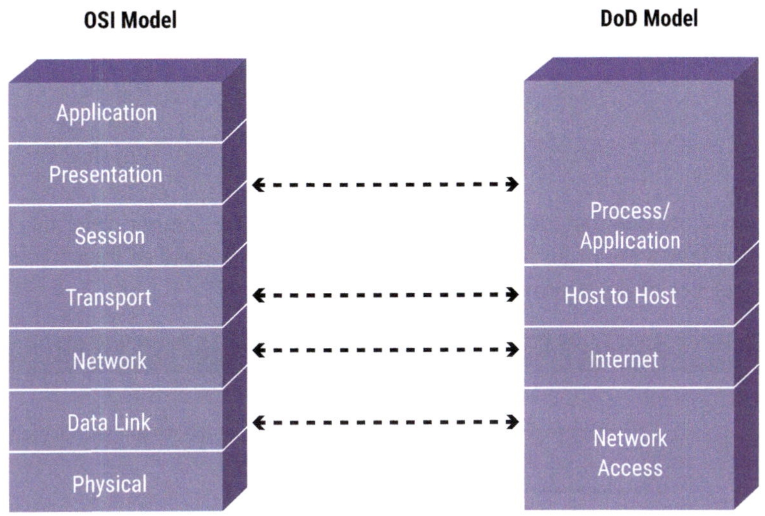

With this disparate work happening, there were many questions to be answered and solutions to be designed to ensure that devices could and would reliably communicate. In the late 1970s, the International Organization for Standardization (ISO) began creating such a standard to define how networks would function; later this became a framework of standards. The ISO was a regimented organization bringing some of the management and coordination that was needed to compile and decide best practices. ISO's work continues through today, still helping to define standards in many different areas.

Out of the initial development came the need to break the entire communication process into smaller parts. Those parts became layers in the overall model. Some of the significant requirements of ISO's OSI model project were to:

- Separate the process of communication into smaller pieces to aid the development, design, and troubleshoot network devices and connections
- Ensure that changes made to one layer of the OSI model do not affect other layers
- Define what functions occur at each layer to aid industry standardization and development
- Allow various types of network hardware and software to communicate

The seven layer OSI model organized the functions of all network protocols. The OSI model is comprised solely of software, even though there is one layer referred to as physical.

The layers of the OSI model are (**figure 2.3**):

Figure 2.3—Layers of OSI Model

Layer Number	Layer Title
7	Application
6	Presentation
5	Session
4	Transport
3	Network
2	Data Link
1	Physical

The layers are numbered to show that they occur in descending order. The number order has become an important hierarchical identifier as the layers are often referred to by their number instead of their name. **Figure 2.4** shows the layers of the OSI model and a brief description of each layer.

Figure 2.4—OSI Model

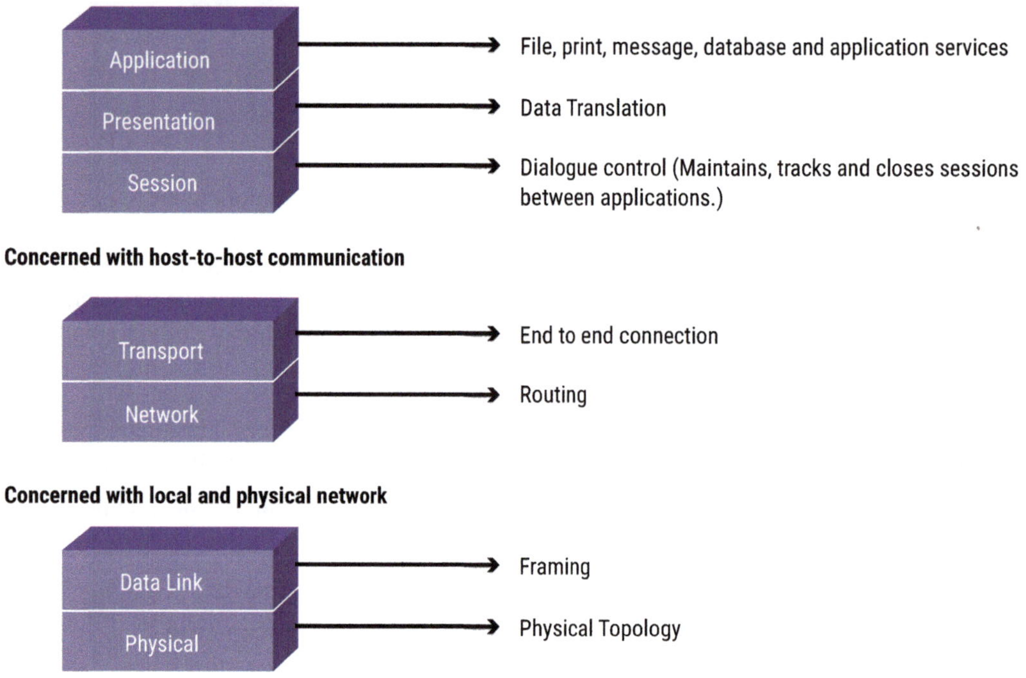

7. Application Layer

The application layer provides a network interface for applications. The application layer supports applications that are network-aware or capable. An email client, for example, can run on a computer and do things such as keep contacts and calendars. The email client won't send any email unless it is connected to a network on a computer that provides the network connections. The application layer provides the bridge between the installed network software and the email client's software to send email.

6. Presentation Layer

The presentation layer translates data to a standard format and manages some of the encryption and data compression. The presentation layer includes protocols like File Transfer Protocol (FTP) and Secure Shell (SSH), developed to facilitate operations like transferring files and securing a connection to another computer over a network. These protocols manage the format of the files and the encryption from computer to computer.

5. Session Layer

The session layer is a collection of protocols that help establish and make applications available on computers across the network. Protocols, like the server message block (SMB) and remote procedure call (RPC), often advertise, answer requests or otherwise make it known that a computer is capable of communicating with specific protocols.

4. Transport Layer

The transport layer manages transmission and verification of data being sent between hosts on the network. Transmitting and verifying can include error control, flow control and establishing and tearing down of the communication connections that are made between hosts. The most well-known protocol at this layer is the Transmission Control Protocol (TCP) which is connection-oriented, meaning it uses most of those listed functions to ensure that every bit of data sent is received and accounted for. In contrast to TCP is User Datagram Protocol (UDP), which is connectionless and sends its data on a best-effort basis assuming the network is reliable enough for the data to successfully reach the destination host.

3. Network Layer

The network layer includes protocols which provide the underlying information required for data to be sent. The network layer uses a logical addressing scheme that makes it possible for hosts on the network to "locate" and be "located" by all other hosts on the network. All hosts must have a logical address to communicate. Providing the information about where the logical addresses are located and how to get there is the responsibility of another set of protocols known as routing protocols. The most widely used network layer protocol today is Internet Protocol (IP), which provides the structure for logical addressing with IP addresses.

2. Data Link Layer

The data link layer includes many complex functions and is divided into two sublayers:

- Media access control layer (MAC)
- Logical link layer (LLC)

The media access control (MAC) sublayer of the data link layer provides an interface to the network adapter which is written into the network interface card (NIC) driver. The NIC driver understands signals from the NIC hardware and learns what the physical medium is (e.g., wired, wireless, optical, etc.). From the physical medium, the NIC driver learns what the physical addressing will be, and how the data is going to be translated and made into the electrical or optical signals. Knowing how the signals are translated lets the NIC driver know how data will be transmitted and what that connection topology to the destination host looks like.

The upper sublayer is called the logical link control (LLC) sublayer; it is responsible for understanding the layer 3 network layer protocol that is being used and how to communicate with that layer. Common protocols at this layer include:

- Ethernet
- Fiber Distributed Data Interface (FDDI)
- Asynchronous Transfer Mode (ATM) for the lower physical cabling types of protocols
- Spanning Tree Protocol (STP)
- Cisco Discovery Protocol (CDP), which helps maintain the topology of the network, share data between devices and helps to prevent errors like loops in the network

1. Physical Layer

Layer 1, the physical layer, helps convert the data stream into electric, optic or analog pulses that cross the transmission medium. These protocols have built-in routines that oversee transmitting data and identifying if there is data loss or corruption. Many protocols exist for each of the different transmission mediums, including EIA/TIA standards for serial cables, 100 and 1000BASE-T connections for Ethernet and many for wireless and optical mediums as well.

2.3.2 Network Topology and Deployments

When discussing network topologies and different deployments, the cables, devices and methodologies have changed significantly over time. Like most things, part of the change was driven by how well things could be made, and how much they cost.

The network topologies are discussed in more detail below. The network topologies are:

- Bus
- Star
- Ring
- Mesh
- Ad hoc wireless
- Multipoint wireless

Bus Topology

Bus topology **(figure 2.5):**

- On the cheaper end of network plant cabling.
- Also known as a linear bus, all devices are connected to a single cable. Also called a trunk or backbone.
- Most often this cable plant was coaxial cable, called Thicknet and Thinnet, depending on the generation and thickness of the cable.
- Computers are connected to the bus using a drop cable, more often called a T-connector or a vampire clamp, as they would clamp onto and pierce the cable rather than cutting the cable and adding a connector.
- Each end of the cable is terminated to prevent the signals from mirroring back down the line.

Figure 2.5—Bus Topology

Star Topology

In star topology (**figure 2.6**):

- Each device connects directly to a network device, such as a hub, bridge or switch.
- Each device has its own cable.

- With a bridge or switch, each device interacts in its own collision domain.
- Higher cost with network devices and more cable.
- Troubleshooting is easier with separate connections to each device.
- Additional network devices can be added to create a hierarchical star.

Figure 2.6—Star Topology

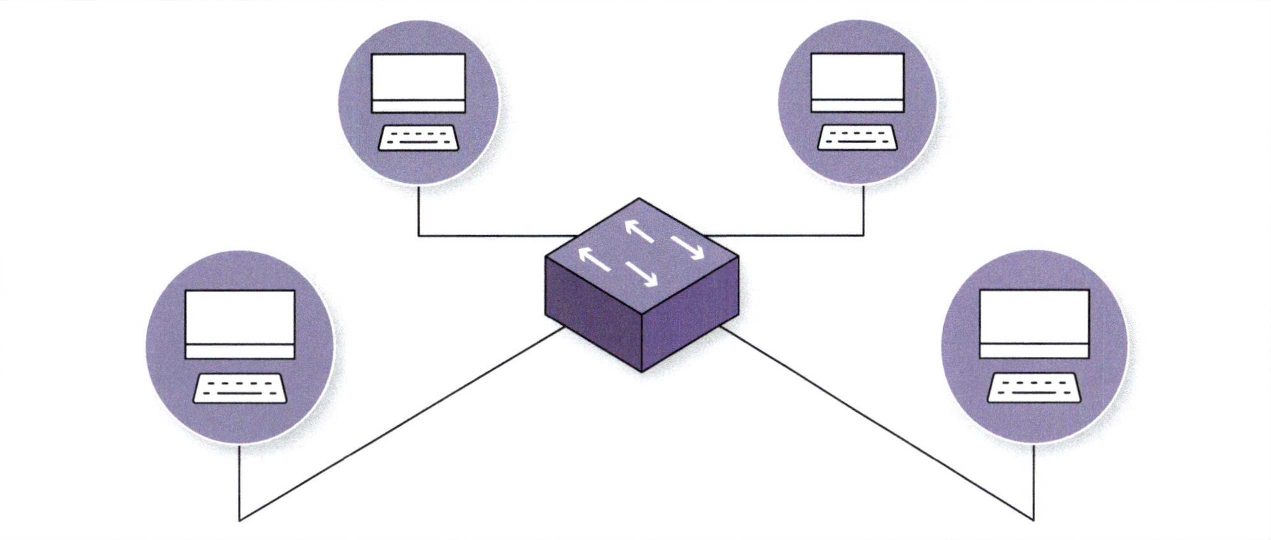

Ring Topology

In ring topology (**figure 2.7**):

- Forms a complete ring, computers can connect from one to the next, or use special network devices.
- Data is passed from one device to the next until it reaches its destination.
- Good mid-use of the amount of cable needed.
- Easy to troubleshoot.
- Faults in the ring can affect the entire network.

Figure 2.7—Ring Topology

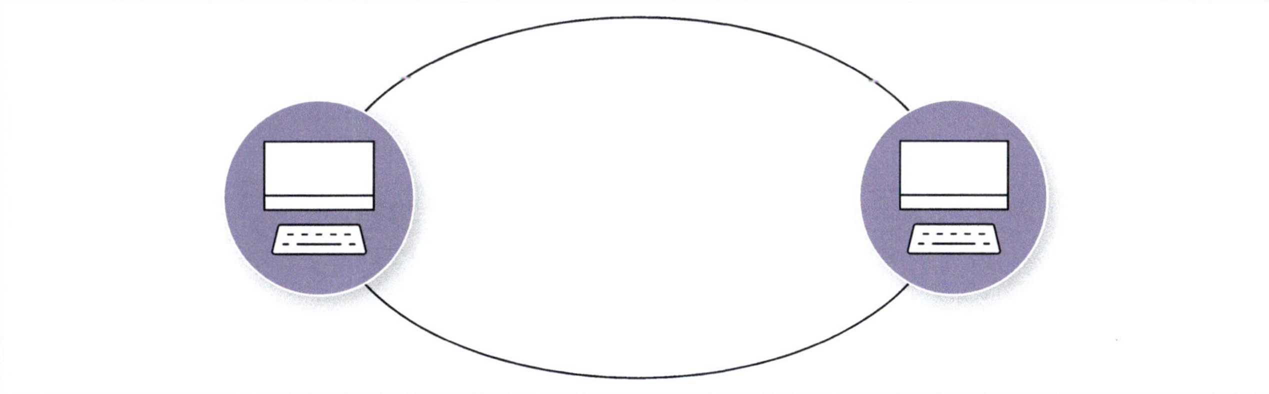

Mesh Topology

In mesh topology (**figure 2.8):**

- Requires that every device on the network be connected to every other device.

- More difficult to implement, high amount of cable, many connections on each device.
- Highest level of redundancy, upgrades and changes can be made with no effect on the rest of the network.
- Easiest topology to troubleshoot.

Figure 2.8—Mesh Topology

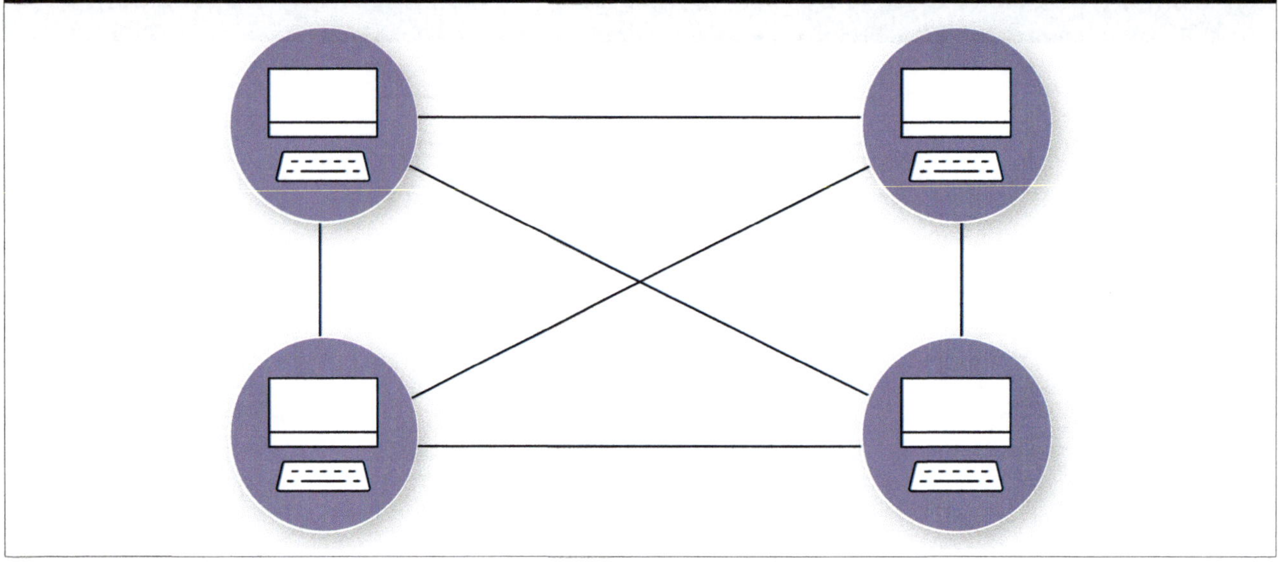

Ad hoc Wireless Topology

In ad hoc wireless topology (**figure 2.9**):

- The physical layout is not inhibited by using wires.
- Networks like this are created when two or more devices with RF transcievers come within range of one another and support ad hoc networking.
- No cost in cabling, only cost is wireless adapter/NIC.
- Troubleshooting and maintaining connections can be difficult, not all ad hoc protocols are the same.

Figure 2.9—Ad hoc Wireless Topology

Multipoint Wireless Topology

In multipoint wireless topology (**figure 2.10**):

- These networks can have many devices attached and each communicates with a wireless bridge.
- A wireless bridge is also called wireless access point (WAP).

- Again, no cost for wires, but there is now a cost for the WAP device.
- Troubleshooting is easier than ad hoc topology.
- All devices must conform to the standard of the WAP.

Figure 2.10—Multipoint Wireless Topology

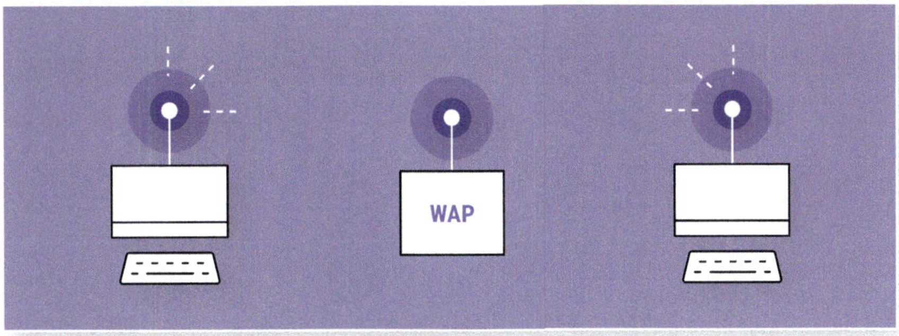

Physical Deployments

When physically deploying a small, basic network, there are additional items that help improve a successful deployment. From a reliability and throughput perspective, a much more robust network can be built without much additional cost. From a troubleshooting perspective, understanding these things can save time. Additional items to help improve successful deployment includes collision and broadcast domains within a network setup.

Collision Domain

Collision domains are network segments with multiple hosts. The collision domains have a shared access medium and effectively share the available bandwidth of the network. If two hosts transmit at the same time, a collision occurs, which means that neither of their data can be received by the intended destination. Collision domains are created using devices such as hubs and WAPs. Additional collision domains can be created with bridge and switch interfaces.

Broadcast Domain

Broadcast domains are the area of the network where all devices in that area can see and participate in broadcasts transmitted from all other hosts and servers in that same area. A routing interface creates broadcast domains on a router, multilayer switch, firewall interface, or some other devices the ability to route information. There are two types of messages:

- Unicast message
- Broadcast message

Unicast Message

A unicast is a message sent from a single computer that has a single destination host. If a computer knows the address of the destination host, a message can be sent directly to the host.

Broadcast Message

However, if the same computer does not know the address of the destination, or there is a message that the host wants all computers to receive, then a broadcast message can be sent. Since a broadcast is a one-to-everyone communication all computers in the broadcast domain receive the message.

The series of four figures below shows the relationship of collision domains and broadcast domains with various device interfaces. **Figure 2.11** is an example of a single hub, with four hosts connected to it. This entire network, also called a local area network (LAN), is a single-collision domain.

Figure 2.11—Single Hub with Four Hosts Connected

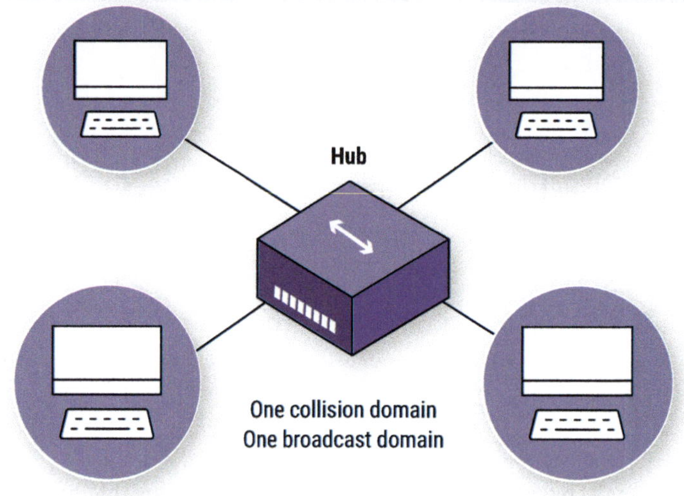

Figure 2.12 shows a bridge, which has the ability to create collision domains from each of its interfaces.

Figure 2.12—Bridge Used to Create Collisions of Interfaces

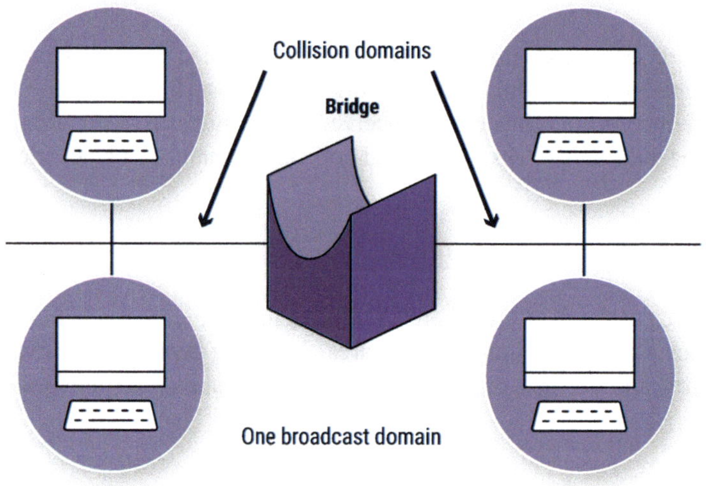

Similar to figure 2.11, **figure 2.13**, illustrates where a switch is used and each of its interfaces create an independent collision domain. As the cost of switches has come down, bridges are being phased out, and switches provide significantly more functionality than bridges.

Figure 2.13—Switch Used to Create Independent Collision Domain

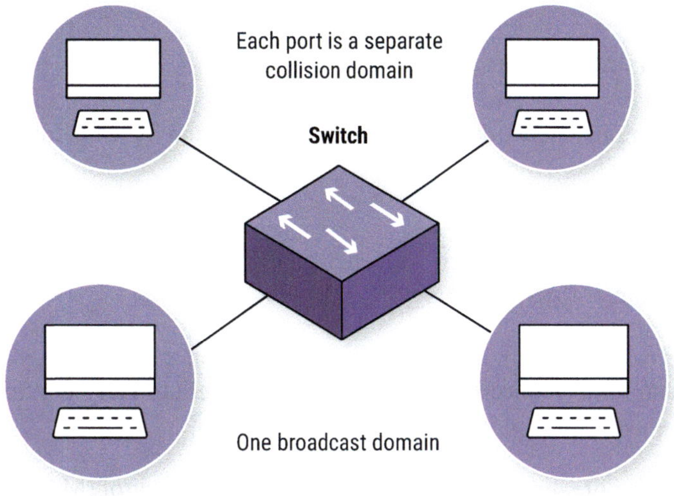

Every interface on a router creates a new broadcast domain, and with every new broadcast domain, there is a new collision domain within it. Routers connect multiple network segments together. A single interface on a router is called a network, network segment, broadcast domain or subnet. On a router with multiple interfaces, each of the interfaces is a separate network segment connected in the router. An Internetwork is comprised of multiple networks connected together.

Figure 2.14 shows an Internetwork with four networks connected to it. Routers are typically the device that connect a network to a wide area network (WAN) or the Internet. There are usually firewalls on one or both sides of the router. Also depicted in the diagram are LAN segments, commonly shown as solid lines and WAN segments shown as dashed lines. The term Internet (world wide web) was shortened from the term internetwork. The Internet is simply the product of every network in the world being connected together by routers to make the largest of Internetworks.

Figure 2.14—Router Creating Broadcast Domains

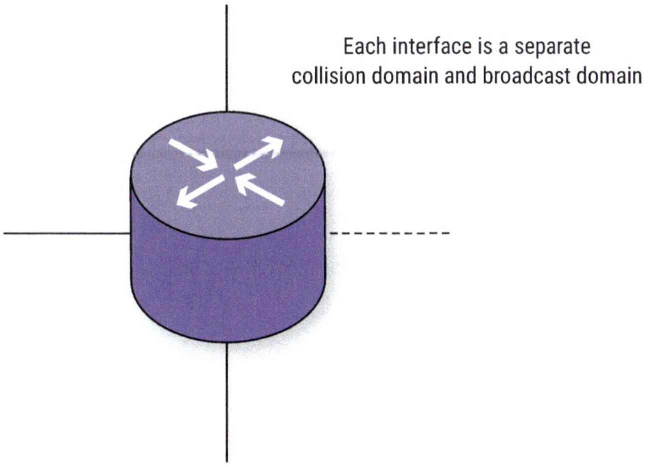

LAN and WAN are generic terms that roughly define a geographic area. A LAN is a local area network, which usually means inside a single office, home or floor in a building. A WAN is a wide area network, and typically the networks that connect multiple offices or buildings together over a larger area. A WAN connection is often purchased through a service provider that owns very large networks that can connect to networks further away.

2.3.3 Network Protocols and Addressing

The main types of addressing within a network are physical addressing and logical addressing.

Physical Addressing

Physical addressing at layer 2—the data link layer—depends on the type of physical network that is used, such as Ethernet, which is the most common LAN technology.

Fiber and optical connections can be used in different ways depending on the technology, such as ATM, or other WAN connections, like T1s and serial connections. Some optical connections carry multiple streams of data at the same time by using different wavelengths of light.

At layer 2, the most common type of physical address is the media access control (MAC) address. The MAC address originally came from the Xerox Network Systems Ethernet address, but is now developed, managed and allocated by the Institute of Electrical and Electronics Engineers (IEEE) as part of the "802" working groups for standards. The IEEE 802 working group has many sub-groups that develop and manage other common standards which also use MAC addresses for physical addressing:

- 802.3 Ethernet,
- 802.11 Wireless LAN
- 802.15 Wireless PAN (personal area network)/Bluetooth

For proper functionality on a network segment, the MAC address must be unique among all hosts. The MAC is assigned to the host's NIC, and the host only uses that address for communications on its local segment.

Format of a MAC Address

A MAC address is a 48-bit address expressed in hexadecimal notation, typically in six groups of two hexadecimal digits (8-bits per grouping). The address can be written without any separating symbols but often written with hyphens or colons between the 8 bit groups. Examples of how each of these notations look are shown below:

XX-XX-XX-XX-XX-XX

XX:XX:XX:XX:XX:XX

XXXX.XXXX.XXXX

MAC addresses fall into one of two different categories based on how they are assigned.:

- Universally administered addresses (UAA)
- Locally administered addresses (LAA)

Universally Administered Addresses

Universally administered addresses (UAA) are administered and assigned to a device NIC by the manufacturer of the device. The IEEE controls and administers the part of the address that identifies the manufacturer. The first 24 bits of the address is called the organizationally unique identifier (OUI). A unique OUI is assigned to each device manufacturer. Multiple unique OUIs may be assigned depending on the amount of equipment the manufacturer makes. Except for the OUI, there is little control over how the manufacturer assigns the addresses, other than requiring the OUIs to be unique (**figure 2.15**).

Figure 2.15—Network Addressing

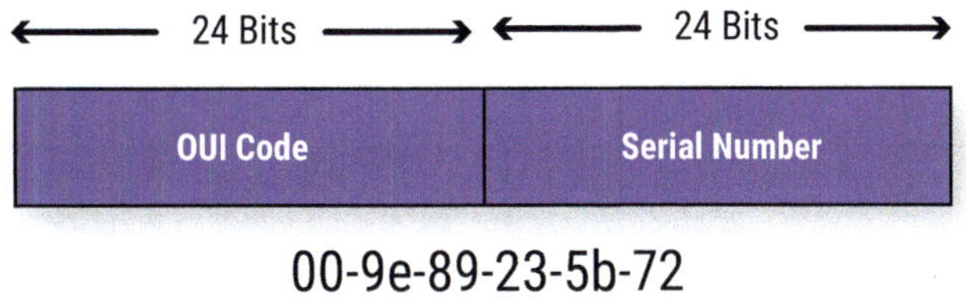

When assigning the MAC addresses onto the NICs, some manufacturers assign or store them in ROM, which means that the MAC address can't be changed. For this reason, these addresses have often been called burned-in addresses or hardware addresses. Not all devices are handled this way—many NICs and interfaces on devices are set up in software, which allows the system administrator to change the addresses if needed.

Locally Administered Addresses

Locally-administered address (LAA) assignment is useful if there are any problems with duplicate MACs on the segment which doesn't happen very often. The more common use for an LAA is so a system administrator can dynamically assign an address if an interface is created or needed for devices that have many interfaces such as virtual interfaces often for switches and servers with hypervisors.

Setting the Type of Address on a Device

The device manufacturer sets the type of address a device will use by setting the universal/local bit (U/L). The U/L is the second least significant bit in the first octet of the address, or reading from left to right, the 7th bit from most significant in the address. To see what that means, refer to **figure 2.16** at the B1 bit.

Figure 2.16—Locally Administered Addresses Scheme

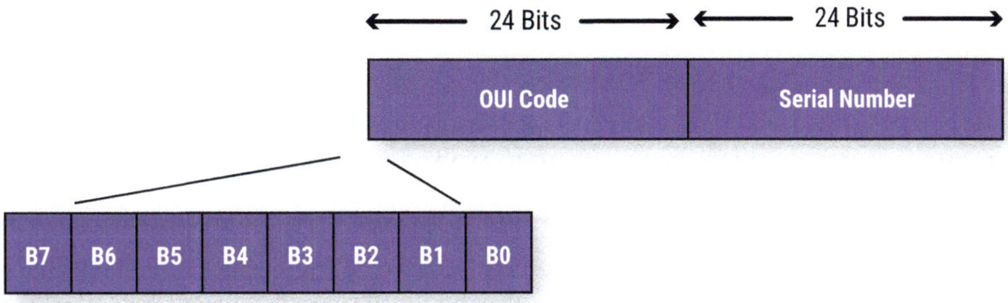

Sending an Entire Ethernet Frame

When an entire Ethernet frame is sent out of a NIC onto the network segment, it has the address of its own NIC (source address) and the address of the host to which the data is being sent (destination address). If the host is sending a broadcast frame to all computers, then the destination address is set to all Fs (FF:FF:FF:FF:FF:FF). Then, every host knows to look at the frame. The structure of the entire frame looks like the frame in **figure 2.17**. Remember: DA is the destination address and SA is the source address.

The preamble lets the devices synchronize and know that a frame is being sent.

- The length field tells the devices how big the frame is.

- The data field holds the data that is being sent from the application.
- The frame check sequence (FCS) is a verification field that lets the destination knows it received the entire frame.

Figure 2.17—Ethernet Frame

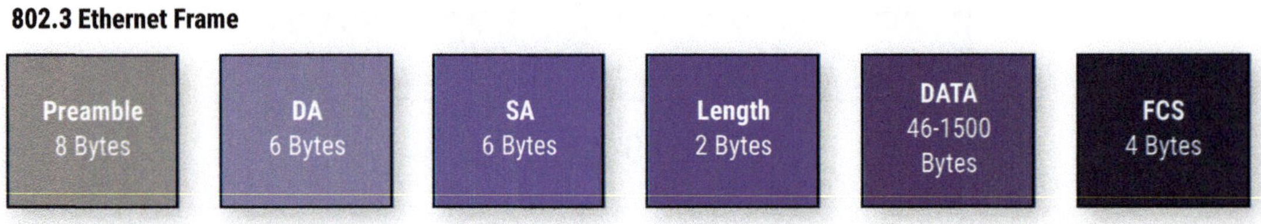

Logical Addressing

Logical addressing happens at layer 3 (network layer) and is determined by the network's protocol suite. In today's networks, the protocol suite is primarily IP. In the earlier days of networking, there were other protocols, like IPX or AppleTalk, but support and development for those protocols has ended.

In layer 3, computer devices are concerned with logical addressing. Logical addressing differs from physical addressing when considering where in the network the address is relevant. Physical addressing is only used within a single network segment. Logical addressing is used end-to-end, regardless of how many network segments the address may be away from the source or destination. This means the same thing whether across a router, across an Internetwork or across the Internet; the source computer will have the logical address of the destination host.

Other logical addressing suites still exist, but they are not used as much on the Internet compared to IP addresses. A logical IP address is structured and works to locate every individual host on the Internetwork.

To send a letter with the destination of 321 Broadcast St., Network, CA 95050, as soon as the letter enters the mail system, the system breaks down that address and starts moving the letter step-by-step closer to its destination.

The zip code and state first tell the system to which big area this letter goes. It needs to be sent by plane if the destination address is across the country. In IP terms that is like being sent out of the local network and put on a big WAN network.

Once the letter gets to a mail hub in CA, the mail system looks at the city, which sends the letter to the nearest mail hub in the city. In IP terms, this would be getting out of the WAN network and finding the network of the company or organization where the destination computer resides.

Now the letter is in the proper city mail hub, and the system looks at the street name, which tells the system which truck the letter goes on to get to the correct street within the city. In IP terms this equals being inside the destination companies internetwork and the street is like getting to the correct router that has the destination network segment connected to it.

Finally, the letter is on the correct delivery truck, that truck is on the correct street, and then the truck goes down the street looking for the specific address of the house or business. This correlates to the specific IP address of the destination host on the network segment where it resides.

IP addresses have had a few versions over time, currently the most widely used version of IP is Internet Protocol version 4 (IPv4).

A history side note—during the years of testing in the ARPANET project, version 1-3 were not formally drafted into standards, so when the Internet Protocol suite was released, the first officially-drafted version was given the version

4 designation. This decision left versions 1-3 in the lab, never becoming actual standards of any kind. Version 5 was written as a separate connection-oriented protocol called Internet Stream Protocol (ST), but was never implemented in any impactful way and later made obsolete. Version 7 was originally written as the next-generation upgrade to IP. Version 7 was used by mistake, because the author thought that the ST-II protocol was using the version 6 designation. The author of ST-II actually reused version 5, leaving version 6 unused. So in the early 1990s, when engineers realized that version 4 would eventually run out of address space with the expansion of the Internet, the outcome of a fully redeveloped Internet Protocol was written and given version 6 designation. IPv6 is discussed in greater detail in ISACA's *Networks and Infrastructure Study Guide*.

IPv4 addresses are hierarchical in their design and framework. The address is 32 bits long and is expressed in what is called the dotted decimal notation. This means the numbers are expressed in decimal (base 10) numbering. The 32 bits are separated into 4 groups (4 bytes) divided by periods, which looks like this:

- 10.0.0.0
- 172.16.0.0
- 192.168.0.0
- 192.0.2.0
- 198.51.100.0
- 203.0.113.0

The network ID portion of the IP address is always the left part of the address. The right part of the address is always the host ID portion. This can be seen in **figure 2.18**.

Figure 2.18—IPv4 Addresses

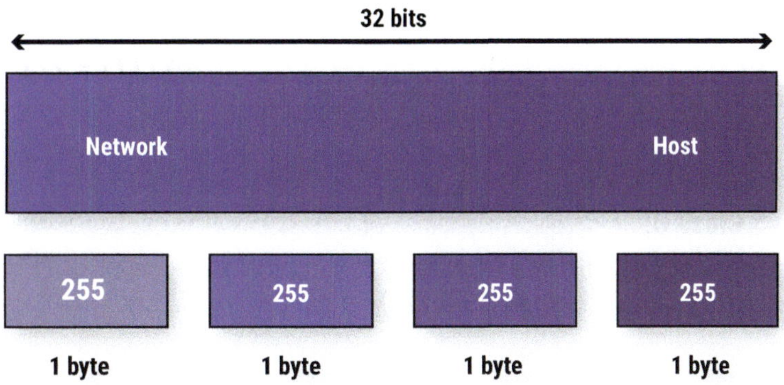

Network addressing is organized by classes based on the number of hosts in the network. The classes are:

Class A: largest for companies and organizations that are the largest based on the number of hosts on their network.

Class B: mid-range for companies that are mid-sized.

Class C: smallest for small companies and maybe even some homes that might get connected. This planning occurred in the late 1970s and early 1980s, before broad use of the Internet and everyone connected at the same time.

Class D: reserved for multicast networks. A multicast network is a group-directed communication, or one-to-many. If a host needs to send data to more than one computer, the host has 3 options:

- Send the data as a unicast and send it 10 times.
- Send the message as a broadcast and interrupt every host on the network.

- Send a multicast message, which does have to be configured since the hosts that want to receive the data, request to be put on the multicast list.

If the hosts are on the multicast list, when the source host has data to send, it will be sent on a multicast address to those requested hosts.

Class E: reserved for research and future use.

Determine the Class of an IP Address

To determine the class of an IP address, look at the first number, which is the left most octet.

If all 8 bits are 0s, then the range starts at 0. $2^0 = 0$

If all 8 bits are 1s, then the range ends at 255. $2^8 = 256$, $256-1 = 255$, minus 1 because counting starts at 0.

So in decimal numbers, for the first octet, the classes breakdown as shown in **figure 2.19**.

Figure 2.19—IP Addresses Class Breakdown

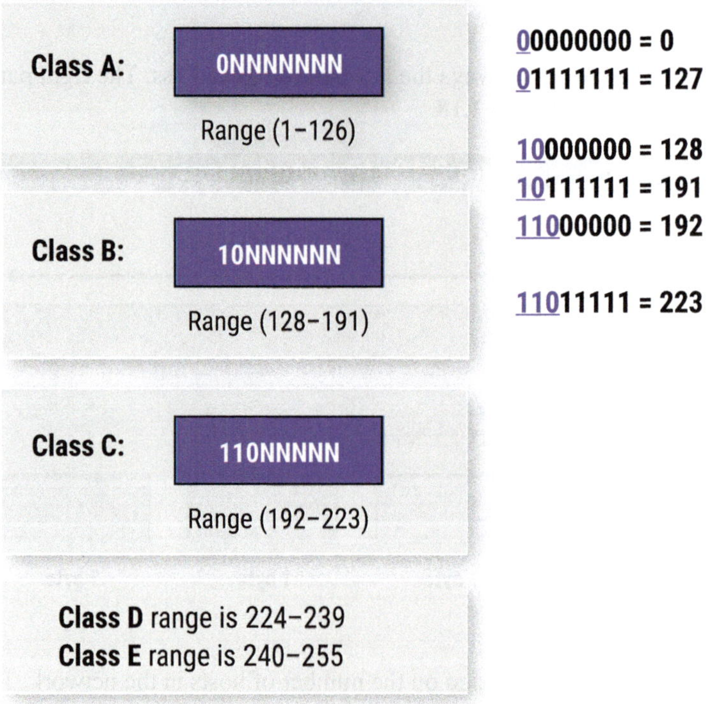

The entire network of 127 was used for something called loopback which allows the host NIC to be tested to ensure its own network software stack is functioning.

Typical binary conversion involves counting up from 0 and then adding the value of each consecutive place holder. It is sometimes easier to work with the values of 8 bits worth of place holders, since the addresses are more often added from the most significant bits rather than the least significant. **Figure 2.20** shows the visualization of this idea.

Figure 2.20—Binary Conversion Table

Bits	Binary	Decimal
0	00000000	0
1	10000000	128
2	11000000	192
3	11100000	224
4	11110000	240
5	11111000	248
6	11111100	252
7	11111110	254
8	11111111	255

Knowing powers of 2 for binary math (**figure 2.21**) makes addressing work and troubleshooting much easier. Here is a quick reminder:

Figure 2.21—Power of 2 for Binary Math

$2^1 = 2$	$2^9 = 512$
$2^2 = 4$	$2^{10} = 1,024$
$2^3 = 8$	$2^{11} = 2,048$
$2^4 = 16$	$2^{12} = 4,096$
$2^5 = 32$	$2^{13} = 8,192$
$2^6 = 64$	$2^{14} = 16,384$
$2^7 = 128$	$2^{15} = 32,768$
$2^8 = 256$	

Understanding IP Classes

Knowing the classes, it is possible to determine how much of the IP address is network, and how much is host address (**figure 2.22**).

Figure 2.22—Network and Host Usage for Different IP Classes

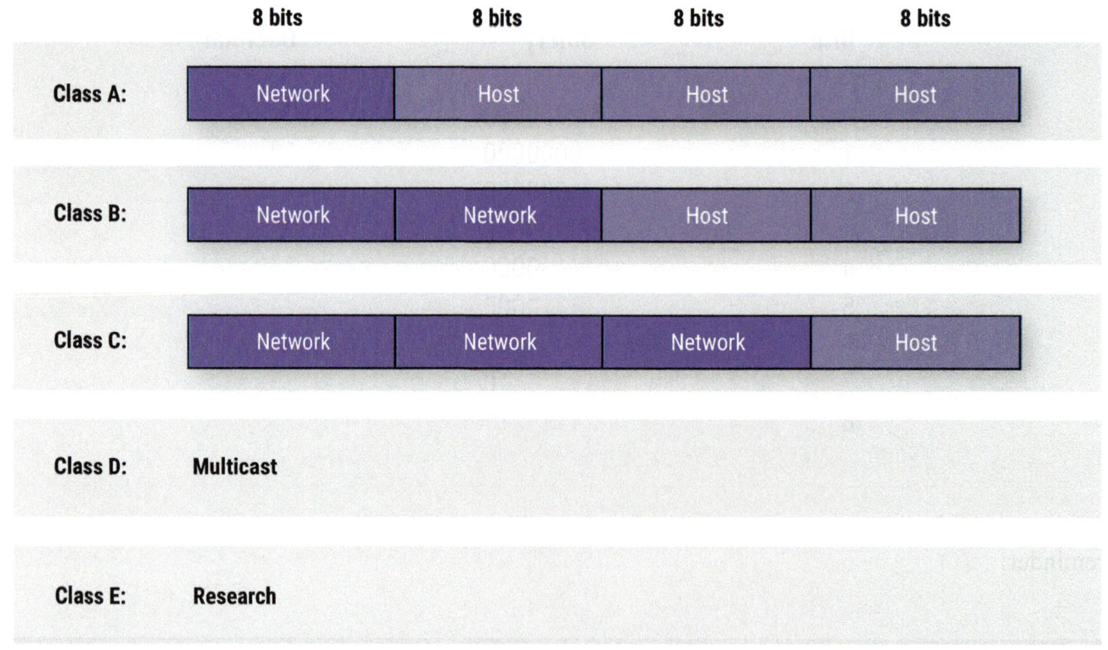

Class A: the first octet is the network ID and the second, third and fourth octets are the host ID.

Class B: the first and second octets are the network ID and the third and fourth octets are the host ID.

Class C: the first, second and third octets are the network ID and the fourth octet is the host ID.

Determine the Number of Hosts on a Network

It is possible to determine the number of hosts on each different size network, as this was the original design theory—to have different networks for large, medium and small organizations.

Class C Network. Starting with a small, Class C network, where the range in the first octet is 192-223, an example from that range would look like this: 192.168.2.0

There are 8 host bits ($2^8 = 256$) allowing possible numbers 0 to 255. However, not all 256 numbers can be used, or assigned to hosts. The first number in the range of host numbers identifies the network ID, and the last number in the range is used as the broadcast address. The broadcast address is the address a host can use to send data to every host on the network. The range looks like this:

Network address:	192.168.2.0
First valid host:	192.168.2.1
Last valid host:	192.168.2.254
Broadcast address:	192.168.2.255

Class B Network. While a Class C network has 254 possible hosts that can be assigned to the network, the range for Class B is 128-191 in the first octet, so an example IP address could be 172.16.0.0.

Since there are 16 host bits in a Class B network, there can be 2^{16} = 65,536 possible addresses. There will always be the same loss of two numbers: the first in the range is used for the network and the last in the range is used for the broadcast. The range from the first number to the last number is a bit different in a Class B network. Since these are binary numbers being represented as decimal numbers, when the fourth octet is full, a 1 is placed in the third octet in a binary place value. Just as with base 10 number counting, when the 1's place is full the next number begins with a 1 in the 10's place. When an octet of numbers is full, it carries over to the preceding octet. When the number places run out in the fourth octet, a single value carries over to the third octet. Here is a range, including the first and last valid hosts, the network address, and broadcast address for an example network:

Network address:	172.16.0.0
First valid host:	172.16.0.1
Last valid host:	172.16.255.254
Broadcast address:	172.16.255.255

There are 65,534 hosts between 172.16.0.1 and 172.16.255.254. Just like a Class C, the range starts from 172.16.0.1 up to 172.16.0.254, however 172.16.0.255 is not the last number in a Class B range. There are no more bits in the fourth octet, and at 255 they are all turned on to 1. The fourth octet is set to 0 and a digit is entered in the third octet; which makes it a 1. So, the next useable host address is 172.16.1.0, which looks like a network address, but is not in this range.

The bits in the fourth octet counting sequence again:

- 172.16.1.1
- 172.16.1.2
- 172.16.1.3
- 172.16.1.4 to 172.16.1.255

When the bits in the fourth octet are full again, another bit carries over to the third octet, so the next address is 172.16.2.0, 172.16.2.1, 172.16.2.2, 172.16.2.3, etc. Bits continue carrying over to the third octet until it is full, at 172.16.255.0, and then up to 172.16.255.254 as the last usable host address, and 172.16.255.255 is finally the last number in the range and the broadcast address.

Class A. Class A works the same way as a class B, but has many more numbers in the host range. Class A range of networks is from 1-126 in the first octet, so an example could be 10.0.0.0. This means there are 24 host bits (2^{24} = 16,777,216 possible addresses). As with Class C and Class B the same two addresses for a network and a broadcast are used.

Network address:	10.0.0.0
First valid host:	10.0.0.1
Last valid host:	10.255.255.254
Broadcast Address	10.255.255.255

If the first host is 10.0.0.1, the range continues to 10.0.0.255. Then, a digit is placed in the third octet and the range is now at 10.0.1.0. Keep counting all the way through to 10.0.255.255, which was the broadcast for a Class B range, but now there is the entire second octet of bits to use. When a Class B is counted up to its full range of 65,536 possible numbers, that is the first of 256 times that will be done with a Class A network range. So after 10.0.255.255, it goes right to 10.1.0.0, then 10.1.0.1, 10.1.0.2 and all the way up to 10.1.255.255. Then, another bit is carried over to the second octet, and 65,536 more hosts can be counted. Bits are carried over into the second octet until it reaches 10.255.255.254, which is the last usable host on the network. 10.255.255.255 is the broadcast.

Determine the Number of Available Networks

The number of available networks can be figured out as well.

Class A is fairly straightforward—the range is 1-126 so there are 126 Class A networks.

For Class B, the range is 128 – 192, which gives a total of 64 numbers in the first octet. For each of those first octet numbers, there is a network combination with every number from the second octet; therefore 64 * 256 = 16,384 Class B networks.

Class C has a range of 192-223 in the first octet which is 32 numbers, but for each of those numbers, there are network combinations with each of the numbers in the second and third octets. That is 16-bits worth for each number, or 32 * 65,536 = 2,097,152 class C networks.

- Class A – 126 networks
- Class B – 16,384 networks
- Class C – 2,097,152 networks

As the address scheme was originally designed, this would be the total number of networks that were available to be assigned to whatever government, corporate or private organizations wanted to be connected to the Internetwork. At the time of its development, there was no reason to think that the Internet was going to become what it has become, or that private entities would be connecting to the same Internetwork. This was a DoD-funded experiment designed primarily for its own use, and today the DoD still owns 13 full Class A networks.

Early network designers knew long before the explosion of the Internet that there was a need to break up the networks into more usable pieces. Class A networks having over 16,000,000 host addresses on a single network was untenable. An Ethernet-based network has a maximum number of devices that can be connected and still be functional. The maximum number is less than 2000 and assumes perfect applications that don't broadcast or waste bandwidth. A Class B with over 16,000 host addresses on a single network was still far more than could be used with most network protocols.

Private IP Addresses

Private IP addresses are reserved for internal use and are nonroutable on the Internet. There are three IPv4 private IP address ranges as follows:

- Class A: 10.0.0.0 to 10.255.255.255
- Class B: 172.16.0.0 to 172.31.255.255
- Class C: 192.168.0.0 to 192.168.255.255

Classful Subnetting

The first solution to the dilemma of too many network addresses was classful subnetting. Using classful subnetting allowed class-based networks to be broken down into usable pieces. If a company had a Class B network, they could break that classful network into 256 Class C networks. This allowed for different buildings, floors and departments within the organization to each be on their own subnet.

If the classful networks were broken into pieces that didn't match the class boundaries, as they were originally designated, then it was impossible to look at an IP address and know for sure which was the home network for the address. For these broken-down pieces to be accurately identified, an additional piece of information was needed for reference. This reference data became known as the subnet mask.

A subnet mask is a 32-bit value that allows the recipient of IP packets to distinguish the network ID portion of the IP address from the host ID portion of the IP address. The network administrator creates a 32-bit subnet mask composed of 1s and 0s. The 1s in the subnet mask represent the positions that refer to the network or subnet addresses. Subnet masks for default classful networks are depicted below:

Class A default: 255.0.0.0

Class B default: 255.255.0.0

Class C default: 255.255.255.0

For example, say the network administrator for a company has been assigned a network ID of 172.16.0.0 with a subnet mask of 255.255.0.0. This means that the entire Class B network is available to use. There are many users with hosts spread over a fairly large campus, including two buildings, each with multiple floors. Each floor has a different department, and each of these departments should be on their own subnet, both for organizational and security purposes. To accomplish what is needed, the network administrator can break up or subnet the network IDs into multiple Class C network IDs.

Subnetting a Classful Network

The ease of subnetting with classful networks is that it only requires the administrator to change the subnet mask. The subnet mask is the information telling the hosts what the subnet is. Beyond that, the most difficult thing is keeping all the subnets organized—where they were assigned, and which hosts have what addresses. For example, the 172.16.0.0 255.255.0.0 network could be changed by adjusting the subnet mask, and now there are 256 Class C networks that look like **figure 2.23**.

Figure 2.23—Subnetting with Classful Network

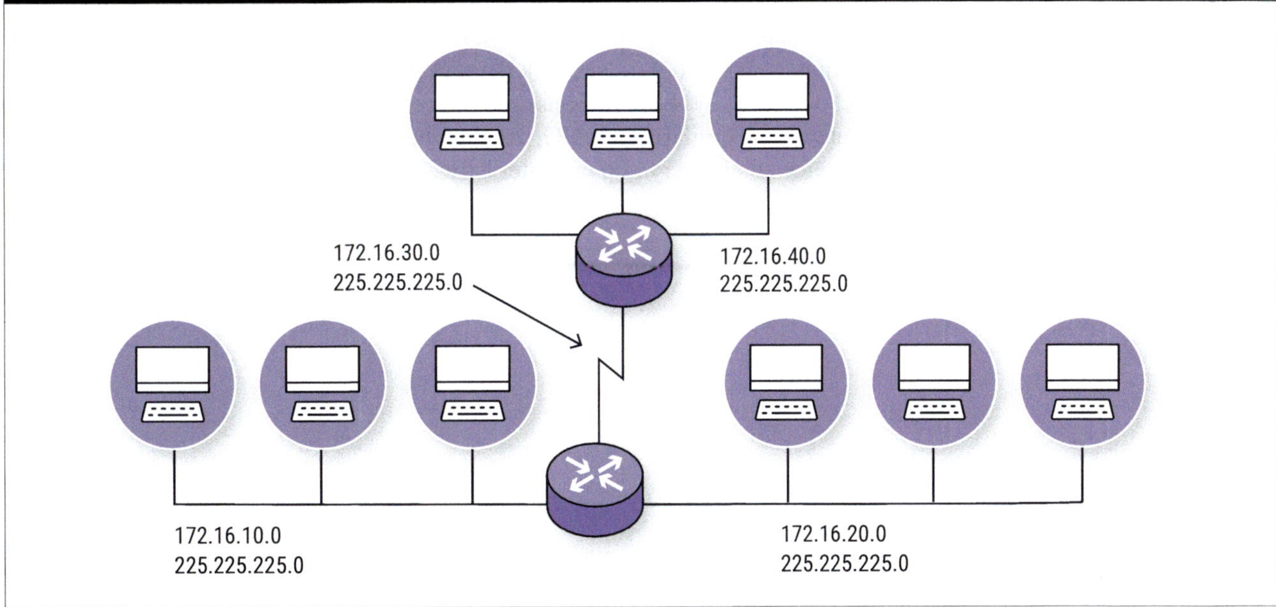

Spacing out the subnets and using every 10th one is a design choice which gives room later to add additional subnets in those areas while preserving the order. All 256 subnetworks are usable, and don't need to be used in order.

Classless Interdomain Routing

Using classful subnetting works in saving address space, but there are still many wasted addresses. It was common to have 100 or less hosts on a subnet that still had 254 host addresses. For this reason, an additional change was made to the addressing scheme, part of a larger change called Classless Inter-domain Routing (CIDR) discussed in RFC4632.[10] The part of this change that dealt with subnetting allowed breaking apart networks into much smaller pieces that don't have to keep the classful network boundaries. Classless subnetting is covered in greater detail in the ISACA *Networks and Infrastructure Study Guide*.

A network device, mostly routers, must have a listing of all subnets on the Internetwork that they can reach with data. A router keeps a listing called a routing table. A routing table has every subnet/network ID listed in it that a router can send traffic to. If a network is not listed in the routing table, then a router cannot forward data to it and drops the traffic. It is possible to manually make sure that every router in an Internetwork has all the subnets listed in the routing table. This would quickly become an administrative nightmare. It would mean that every time there was a change in the network, the administrator would have to tell every router about the change.

Routing protocols can handle this problem. A routing protocol can take a change in the network and dynamically send that new data to all routers. Each of the routers updates the routing tables and begins sending data to the new subnet. Examples of routing protocols are:

- Routing Information Protocol (RIP)
- Open Shortest Path First (OSPF)
- Enhanced Interior Gateway Routing Protocol (EIGRP)
- Border Gateway Protocol (BGP)

Routing protocols are covered in greater detail in the ISACA *Networks and Infrastructure Study Guide*.

Internet Control Message Protocol (ICMP) is a very useful protocol within the TCP/IP suite that is used by all devices to send and receive error messages and operational messages between devices. The most well known thing about ICMP is that both the PING (some say it stands for Packet InterNet Groper) and the traceroute programs use ICMP messages to test connectivity between hosts. PING, much like a submarine SONAR system, bounces back and forth to measure distance or see if something is in front of it. Network devices use PING to test if a destination is alive on the other end. PING uses the echo request and echo response messages within the ICMP protocol to complete the test (**figure 2.24**). Traceroute works similarly to PING, except where PING only tells the device if the destination is there, traceroute tells the device about every hop or router between the source and destination.

[10] Li, T.; "Classless Inter-domain Routing (CIDR): The Internet Address Assignment and Aggregation Plan," The Internet Society, 2006, https://tools.ietf.org/html/rfc4632

Figure 2.24—Ping Sequence

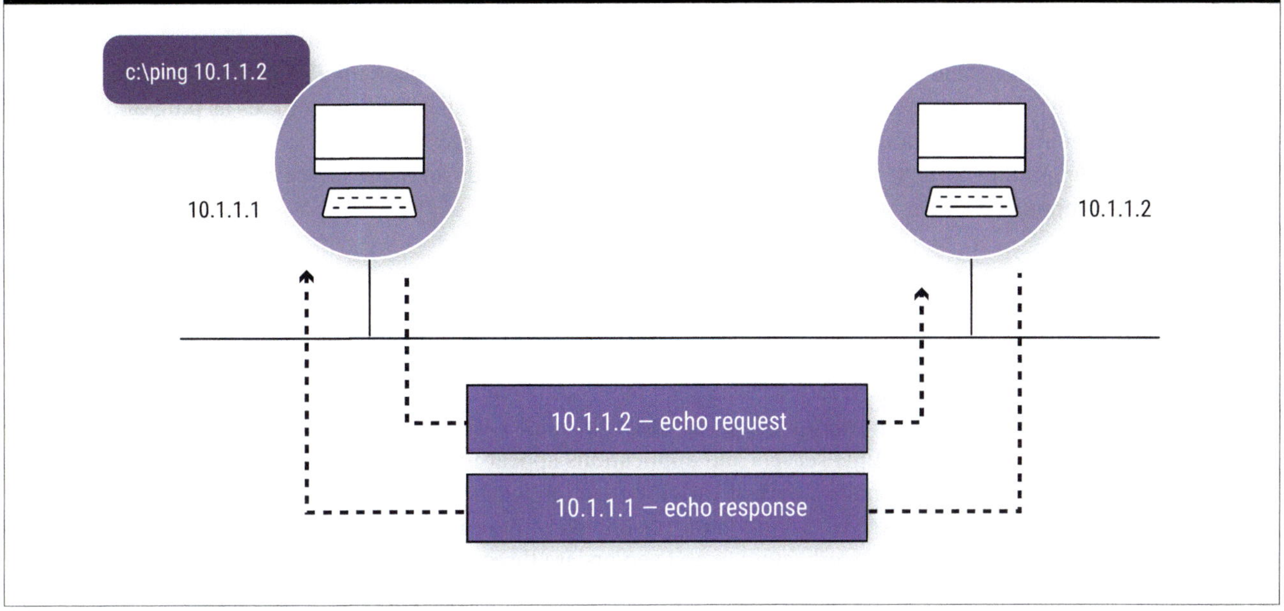

Address Resolution Protocol (ARP) is used to discover a destination host's MAC when only the IP address is known (**figure 2.25**).

Figure 2.25—Address Resolution Protocol

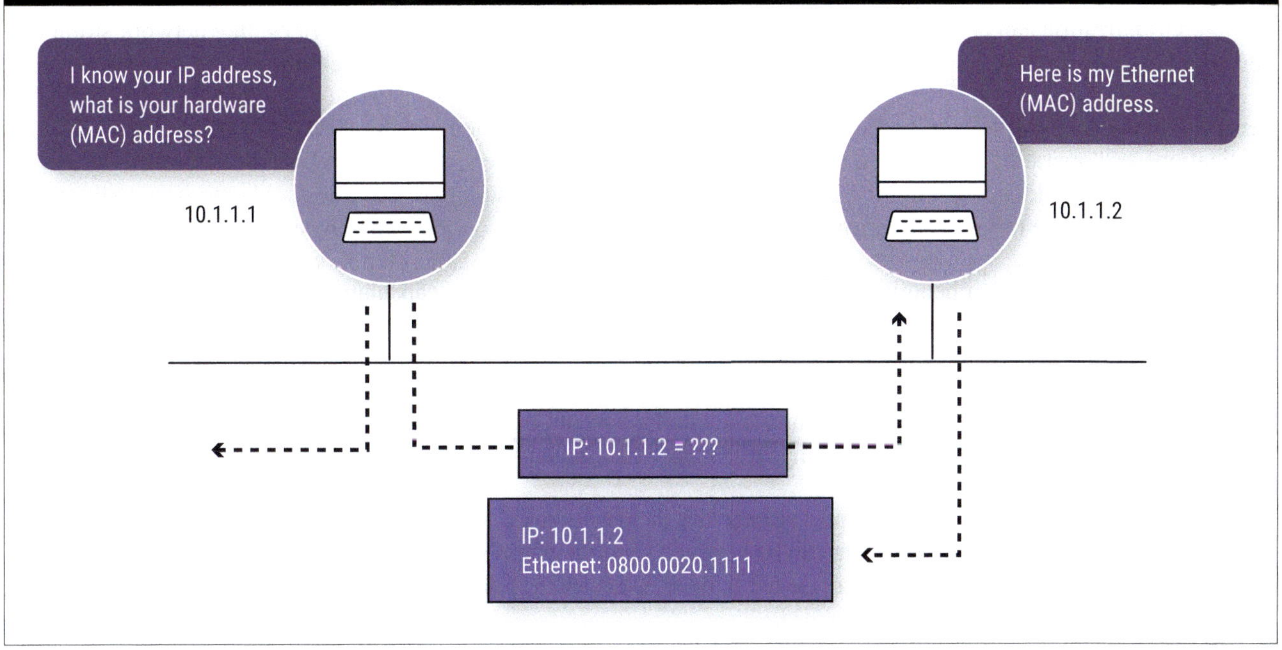

2.4 Virtualization Concepts

Virtualization can hide or abstract resources such as network, storage and CPU from server OSs and applications. The resources appear and perform with more power and speed than they would otherwise. It accomplishes this by concealing the physical characteristics of a resource from the other systems interacting with it.

Before virtualization, every network application, database and authentication server needed its own physical host. Virtualization gives the unique ability to combine many resources into a large available pool. In this way, fewer physical hosts and physical resources may be required to run the same amount of applications. Those resources can be used by virtual machines and applications as they are needed, and released when they're not needed. This can be accomplished because there is a higher-level virtual application looking at all of these resources as plug-and-play components. The goal of virtualization is to reduce cost by keeping fewer physical devices, which can reduce power, maintenance and upgrading costs. Efficiency is another important gain and can mean fewer work hours for technicians and support.

2.4.1 Client-Server Architecture

Virtualization is generally believed to have originated in the late 1960s or 1970s,[11] however it was not until the early 2000s that VMware® grew in popularity for endpoint solutions.[12] An area most affected by the technology has been client-server environments. A super power of virtualization is the ability to make individual computer resources look like multiple machines, or make multiple servers and applications appear to the rest of the network to be a single application. Client-server environments are perfect for virtualization technologies. Physical client resources can be underutilized, but a single physical server running multiple clients can use less resources and make all the same applications available. Virtualizing network, storage and servers are three major objectives of virtualization.

Network virtualization can take what would be multiple connections to many different server hosts and combine them into fewer channels. To the virtual machines, there is no difference in their connection to the network or their ability to access resources. Network resources can be dynamically added and taken away depending on what the need and load is at the time.

Storage virtualization can take physical storage from many different physical storage devices, like network storage devices or storage area networks (SAN), and make all of the storage look like one large pool.

Server virtualization can be configured for applications and operating systems. After a server is virtualized, applications and operating systems can be moved between physical servers. This allows for failover, or immediate maintenance and workload continuity. Server virtualization can also ensure that resources are being used to their fullest potential all the time because virtual devices can run side-by-side on the same physical machine.

Virtualization provides other benefits, including partitioning, isolation and encapsulation. Encapsulation makes moving and backing up a virtual machine easy. From the perspective of the virtualization software, the entire virtual machine, including the operating system, is a single file.

From a security standpoint, isolation can control data and keep data from moving between systems. Isolation also helps prevent any instability in one virtual machine from affecting any of the other virtual machines.

Partitioning has always meant dividing up a storage device which could keep an operating system separate from stored data. In virtualization, partitioning can be between separate operating systems and applications and allows them to run separately or together to increase the scale of users and applications they can serve.

Generally speaking, virtualization provides consolidation. However, encapsulation allows entire business environments to be saved on a few files which are easy to copy and control. If an operating system is virtualized, including its applications and configurations, it can be moved anywhere in the organization without disrupting business continuity. Virtual machines (VMs) maximize availability by automatically migrating from weak or overwhelmed hosts to another virtualized platform.

[11] Oracle, "Brief History of Virtualization," https://docs.oracle.com/cd/E26996_01/E18549/html/VMUSG1010.html
[12] VMware, "VMware Timeline," https://www.vmware.com/timeline.html

Client-Server Virtualization

Client-Server architecture lends itself to the virtualization of applications. Application virtualization, also called Virtual Desktop Infrastructure (VDI), takes the application from user machines and moves it to a centralized server. The experience for the end user is the same because application virtualization addresses the limitations of client-server technology. For example, in traditional client-server technology, client programs must be installed on each user system, therefore client applications must be duplicated for each user. If the client program is virtualized, it is running on a server in a data center instead of taking up space and power on each user system.

Consolidating IT application virtualization in a central data center, can provide other benefits in application delivery. For example, initial application deployment, updates and patches can occur in a matter of minutes or hours. This is in direct contrast to the lengthy investment of manpower that was previously required to make sure enterprise-wide applications were updated and working. Consolidating applications in a data center and making the client virtual also improves security. Application files are not spread across the organization on many user systems which means they can be monitored and maintained more easily.

Client virtualization also increases mobility and flexibility. With the client applications running from a central server, the client applications can be accessed from anywhere and not just the user's PC. Any device that has authorized access to the data center can have access to the application.

2.4.2 Cloud Computing

The history of the cloud goes back to the late 1970s in the ARPANET project, where a cloud symbol was used to represent devices and networks. The word "cloud" in the IT industry has always been used as a metaphor for part of the network where the details did not matter in a particular discussion. The current use of the term cloud computing means the delivery of computing services, over the Internet, using servers, storage, databases, networking or software. The cloud is just another way of saying, "somewhere else, where the details don't matter to me." The promise of faster innovation, flexible resources and economies of scale at a better price is tempting, and the right decision for many organizations. Users typically pay only for cloud services they use, helping them lower operating costs, run infrastructure more efficiently and scale as business needs change. If an administrator is not aware of how cloud cost models work, the benefit of cost savings can quickly disappear.

Cloud computing is a fundamental shift in the traditional way businesses think about IT. The following are some typical reasons organizations may use cloud computing services:

- **Cost**—Cloud computing can eliminate much of the capital expense of buying hardware and software—especially in setting up and running on-site data centers. Rows of racks full of servers requiring electricity for power and cooling as well as IT experts to manage the infrastructure are costly. As with most things that have competition, many of the companies that support and sell having onsite data centers have worked to adjust their cost structure to make it more appealing to keep some of those server resources onsite.
- **Speed**—Most cloud computing services are provided on-demand and with no interaction from the hosting company. Whatever computing resources are needed can be provisioned in just few minutes. This can give a lot of flexibility when resources are needed and eliminates waiting for capacity planning of on-site resources.
- **Scale**—The benefits of cloud computing services include the ability to scale elastically, which is cloud-speak for getting the right amount of whatever resource (computing power, storage, bandwidth), right when and where it is needed.
- **Productivity**—Cloud computing removes the need for many menial tasks, such as racking and stacking, hardware setup and software patching. IT teams can spend time on more important business goals.
- **Performance**—The biggest cloud computing services run on a worldwide network of secure data centers which are regularly upgraded to the latest generation of fast and efficient computing hardware. This offers several benefits over a single corporate data center, including reduced network latency for applications and greater economies of scale.

- **Reliability**—Cloud computing makes data backup, disaster recovery and business continuity easier and less expensive because data can be mirrored at multiple redundant sites on the cloud provider's network.
- **Security**—Many cloud providers offer a broad set of policies, technologies and controls that strengthen security postures, helping protect data, apps, and infrastructure from potential threats.

Types of Cloud Computing

Not all clouds are the same and not one type of cloud computing is right for everyone. Several different models types, and services have evolved to help offer the right solution for different needs.

First, it is important to determine the type of cloud deployment, or cloud computing architecture, where cloud services will be implemented. There are three different ways to deploy cloud services: on a public cloud, private cloud or hybrid cloud.

Public clouds are owned and operated by third-party cloud service providers, who deliver their computing resources, like servers and storage, over the Internet. With a public cloud, all hardware, software and other supporting infrastructure is owned and managed by the cloud provider. Access to these services and account management is done using a web browser.

Private cloud refers to cloud computing resources used exclusively by a single business or organization. A private cloud can be physically located on the company's on-site data center. Some companies also pay third-party service providers to host their private cloud. On a private cloud, the services and infrastructure are maintained on a private network.

Hybrid cloud combines public and private clouds, bound together by technology that allows data and applications to be shared between them. By allowing data and applications to move between private and public clouds, a hybrid cloud gives a business greater flexibility, more deployment options and helps optimize existing infrastructure, security and compliance.

Types of Cloud Services

The following are types of cloud services:

- **Infrastructure as a service (IaaS)**—The most basic category of cloud computing services. With IaaS, IT infrastructure is rented, including servers and virtual machines (VMs), storage, networks, operating systems, from a cloud provider on a pay-as-you-go basis.
- **Platform as a service (PaaS)**—Platform as a service refers to cloud computing services that supply an on-demand environment for developing, testing, delivering and managing software applications. PaaS is designed to make it easier for developers to quickly create web or mobile apps, without worrying about setting up or managing the underlying infrastructure of servers, storage, network and databases needed for development.
- **Software as a service (SaaS)**—Software as a service is a method for delivering software applications over the Internet, on demand and, typically, on a subscription basis. With SaaS, cloud providers host and manage the software application and underlying infrastructure, and handle any maintenance, like software upgrades and security patching. Users connect to the application over the Internet, usually with a web browser on their phone, tablet or PC.

Online services are used to send email, edit documents, watch movies or TV, listen to music, play games, or store pictures and other files—it is likely that cloud computing is making it all possible behind the scenes. Here are some additional services that can be done in the cloud:

- Create cloud-native applications
- Test and build applications

- Store, back up and recover data
- Analyze data
- Stream audio and video
- Embed intelligence
- Deliver software on demand

2.5 Security Concepts

Everyone in an organization from the top leadership down to every single worker and user must actively think about security. Emerging risk and challenges are constantly evolving and everyone must be prepared to help in identifying the risk. Security is more than the protection of devices and assets—it has to include information and data. Within IT, all disciplines, from software development, network, applications and help desk must all work together to combat these challenges.

2.5.1 Information Security Objectives

Information security is a subset of cybersecurity. The goal of information security is to keep data safe from unauthorized access, modification and theft during processing, storage and transmission. Information security deals with all formats of information—paper documents, digital assets, intellectual property in peoples' minds and verbal and visual communications. This differs from cybersecurity, which is the protection of information assets by addressing threats to information processed, stored and transported by internetworked information systems. A comparison of the two is illustrated in **figure 2.26**.

Figure 2.26—Information Security vs. Cybersecurity

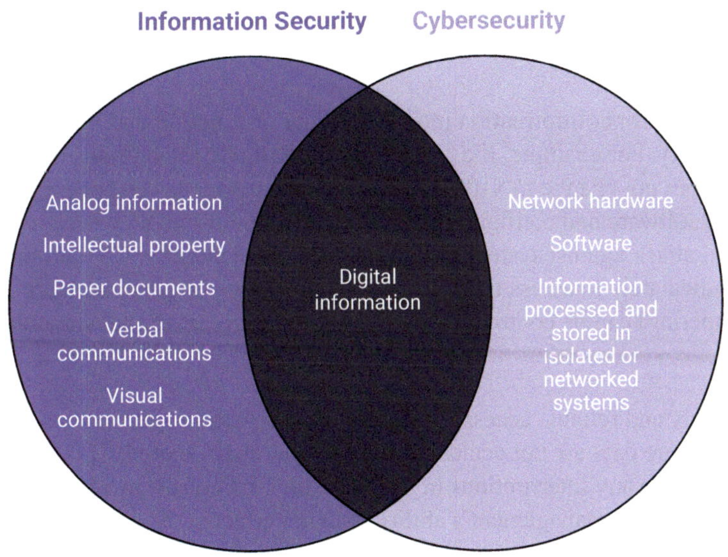

Source: ISACA, *Cybersecurity Fundamentals Study Guide, 3rd Edition*, USA, 2021

The objective of information security is threefold, involving the critical components of confidentiality, integrity and availability—often referred to as the C-I-A triad (**figure 2.27**). Confidentiality means protection from unauthorized access, while integrity means protection from unauthorized modification and availability means protection from disruptions in access.

Figure 2.27—CIA Triad

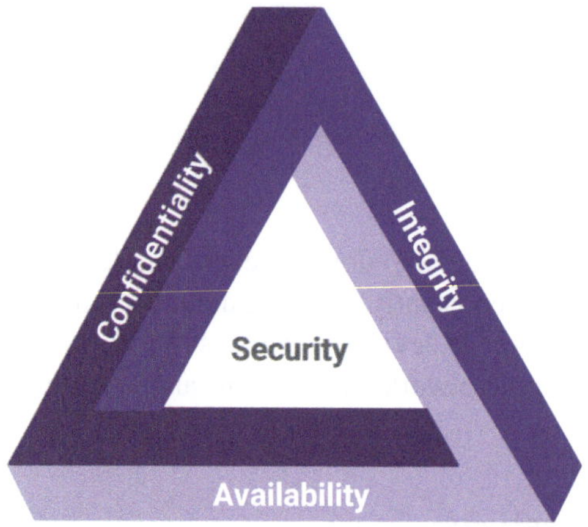

Source: ISACA, *Cybersecurity Fundamentals Study Guide, 3rd Edition*, USA, 2021, https://www.isaca.org/bookstore/bookstore-misc-ebook/witasb

Confidentiality is preserving authorized restrictions on access or disclosure, including means for protecting privacy and proprietary information. Different types of information require different levels of confidentiality, and the need for confidentiality, can change over time. Personal, financial and medical information require a higher degree of confidentiality than the minutes of a staff meeting, for example. Similarly, some enterprises need to protect information on competitive future products before their release but may need to make the same information public afterward. Data must be protected from improper disclosure according to its sensitivity and applicable legal requirements. The confidentiality of digital information can be maintained using several means, including access controls, file permissions and encryption.

Integrity is guarding against improper information modification or destruction and includes ensuring information nonrepudiation and authenticity. For example, if a bank transfers $10,000 to another financial institution, it is important that the amount does not change to $100,000 during the exchange. The concept of integrity also applies to electronic messaging, files, software and configurations. Any violation of integrity is significant, because it may be the first step in a successful attack against system availability or confidentiality. Contaminated systems and corrupted data must be dealt with immediately to assess the potential for further violation or damage. The integrity of digital assets (to include data tampering) can be controlled and verified by logging, digital signatures, hashes, encryption and access controls.

Availability ensures the timely and reliable access to and use of information and systems to the right people. This includes safeguards to make sure data are not accidentally or maliciously deleted. This is particularly important with a mission-critical system because any interruptions in its availability can result in a loss of productivity and revenue. Similarly, the loss of data can affect management's ability to make effective decisions and responses. Availability can be protected by using redundancy and backups, and implementing business continuity management and planning.

2.5.2 Threat Landscape

Every organization is different in what it does, produces or serves; accordingly, risk differs from organization to organization as well. Risk tolerance is influenced by the amount of risk enterprise leadership is willing to incur based on criticality of assets. An asset is something of either tangible or intangible value that is worth protecting, including people, information, infrastructure, finances and reputation.

It is important that all IT personnel understand risk—especially in the context of their computing environments. After all, any change to a computing device or network may undermine existing security measures.

Risk can be defined as the combination of the probability (or likelihood) of an event and its impact.[13] Risk is mitigated through the use of controls or safeguards. The probability of an event is the likelihood that a given threat will exploit an exposed vulnerability. If there are no potential consequences or impact, there is considered to be no risk. Conversely, the greater the consequences or impact, the greater the risk.[14]

A threat is anything (e.g., object, substance, or human) that is capable of acting against an asset in a manner that can result in harm. ISO/IEC 13335 defines a threat broadly as a potential cause of an unwanted incident. Some enterprises make a further distinction between a threat source and a threat event, classifying a threat source as the actual process or agent attempting to cause harm, and a threat event as the result or outcome of a threat agent's malicious activity.

Threat events are caused by threat sources. A threat source is characterized as:

- The intent and method targeted at the exploitation of a vulnerability; or
- A situation and method that may accidentally exploit a vulnerability

In general, types of threat sources include:

- Hostile cyber or physical attacks
- Human errors of omission or commission
- Structural failures of enterprise-controlled resources (that is,

  hardware, software, environmental controls)
- Natural and man-made disasters, accidents and failures beyond the control of the enterprise

Based on the risk to an organization, there are many types of specific attacks that can be carried out against its people and infrastructure—both physical and logical. Below are some of the most common types of attacks:

- **Backdoor**—A means of regaining access to a compromised system by installing software or configuring existing software to enable remote access under attacker-defined conditions. Note that backdoors may also be manufacturer-enabled, with IoT devices.
- **Brute force attack**—An attack made by trying all possible combinations of passwords or encryption keys until the correct one is found.
- **Buffer overflow**—Occurs when a program or process tries to store more data in a buffer (temporary data storage area) than it was intended to hold. Since buffers are created to contain a finite amount of data, the extra information—which has to go somewhere—can overflow into adjacent buffers, corrupting or overwriting the valid data held in them. Although it may occur accidentally through programming error, buffer overflow is an increasingly common type of security attack on data integrity. In buffer overflow attacks, the extra data may contain codes type of security attack on data integrity.
- **Covert channel**—Means of illicitly transferring information between systems using existing infrastructure. Covert channels are simple, stealthy attacks that often go undetected.
- **Denial of Service (DoS) attack**—An assault on a service from a single source that floods it with so many requests that it becomes overwhelmed and is either stopped completely or operates at a significantly reduced rate.

[13] ISACA Glossary, www.isaca.org/resources/glossary
[14] ISACA, *CISM Review Manual*, 15th Edition, USA, 2019, https://www.isaca.org/bookstore/cism-exam-resources/cm15ed

- **Man-in-the-middle attack**—An attack strategy in which the attacker intercepts the communication stream between two parts of the victim system and then replaces the traffic between the two components with the intruder's own, eventually assuming control of the communication.
- **Social engineering**—An attack based on deceiving users or administrators at the target site into revealing confidential or sensitive information. When successful, social engineering can permit unauthorized access to systems. It involves a "con game" that tricks others into divulging information or opening malicious software or programs.
- **Phishing**—A type of email attack that attempts to convince a user that the originator is genuine, but with the intention of obtaining information for use in social engineering.
- **Side-channel attack**—An attack enabled by leakage of information from a physical cryptosystem. Characteristics that could be exploited in a side-channel attack include timing, power consumption and electromagnetic and acoustic emissions.
- **Spearphishing**—A phishing attack designed to entice specific individuals or groups to obtain important information.
- **Spoofing**—Faking the sending address of a transmission in order to gain illegal entry into a secure system.
- **Zero-day exploit**—A vulnerability that is exploited before the software creator/vendor is aware of its existence. Zero-day may also refer to known flaws which do not have a patch available.

2.5.3 Securing Resources and Assets

Securing an organization and all of its resources and assets involves more than just blocking attacks. An organization is made up of physical devices, the information on those physical devices and the people using those physical devices. All of these items must be secured, and the best way to do that is through a layered security approach.

To physically secure people and access to devices, the minimum physical security requirements of using door locks, alarm systems and a security guard is no longer enough. Additional safeguards are required to protect valuable devices and products that may be within facilities. Some of those additional safeguards are:

- **Emergency power**—Improper equipment shutdowns and power surges damage sensitive electronic equipment. Backup power sources are sized for intended system load which requires an understanding of core systems and service needs during power outages. Emergency power systems may include batteries for short term power needs and providing clean power to equipment. Longer duration power outages may require using generators or redundant power connections. Also, to maintain the constant power feed when switching between primary power, automatic transfer switches (ATS) and similar devices may be required.
- **Electromagnetic Interference (EMI) Shielding**—EMI shielding prevents data leakage and interference. Additionally, modern electrical code typically addresses this by specifying how to run network communications cables in buildings.
- **Fire suppression**—modern building codes require fire suppression systems for life safety. Types and locations of fire suppression systems should be considered during business continuity planning.
- **Infrastructure**—server rooms and data centers are typically occupied, if only periodically. However, distributed networking equipment (i.e., wireless access points) and cabling runs require periodic checks. Business continuity planning (BCP) must also account for underground distribution interruptions (i.e., fiber) due to construction. In organizations where fault tolerance and uptime is highly important, the requirement for redundant, geographically-distinct cable paths may be a requirement.
- **Locks**—locks may be traditional types or electronic (i.e., personal identification number (PIN) codes, proximity) or biometric. The type and style of lock is influenced by the sensitivity of the information processing facility.
- **Airlock (aka mantrap)**—an access control mechanism used for accessing sensitive areas. Mantraps have locked doors on both sides with a lock mechanism closest to the sensitive space. Failure to successfully enter an access code (may be keycode, proximity reader, biometrics) locks the person in the room pending a human response.

- **Proximity Readers**—these are access cards that are read from a short distance away via electromagnetic waves. Primarily read with LF signals at 125 kilohertz, which can also be used for basic location tracking of valuable personnel and property.
- **Real-time location tracking (RTLS)**—Similar to proximity and passive readers, this technology can be used with active tracking devices providing updated location information of valuable personnel and property, as well as, inventory control for extremely valuable property. In some cases these systems are used for evacuation monitoring and control in hazardous or very secure locations.
- **Temperature and Humidity Controls**—environmental controls are increasingly connected to IT networks and the Target breach is a popular reminder of the threat external access allows.

Technical controls are IT-specific software, tools and technologies that help minimize cyberrisk. Cybersecurity is a dynamic and ever-changing environment and requires continuous monitoring, updating, testing, patching and changing as technology and business evolve. These controls are critical to maintaining security within any organization's IT infrastructure. Failure to address these processes is one of the top causes of security breaches in organizations. Examples include:

- Antivirus software
- Data loss prevention (DLP) solutions
- Encryption
- Firewalls

Administrative controls address the human element and therefore are occasionally called management controls. Administrative controls include:

- Policies
- Procedures
- Guidance
- Security Awareness Training

2.5.4 Security Controls

The following are common security controls:

- **Identity and Access Management (IAM)**—IAM encapsulates people, processes and products to identify and manage the data used in an information system to authenticate users and grant or deny access rights to data and system resources. The goal of IAM is to provide appropriate access to enterprise resources. IAM relies on access control to carry out who can do what on information systems. Access control consists of identification, authentication and authorization. Another key aspect of access control is accounting. An industry framework to control access is often referred to as AAA—authentication, authorization and accounting.
- **Identification**—Cybersecurity relies upon the establishment and maintenance of user profiles that define the authentication, authorization and access controls for each user. Today, organizations have a variety of ad hoc processes and tools to manage and provision user identity information. Identity management focuses on streamlining various business processes needed to manage all forms of identities in an organization—from enrollment to retirement. Usernames remain the prominent method of user identification.
- **Authentication**—Authentication is the process of proving that people are who they say they are. This procedure is at the core of every security system in use today. These systems have layers of protection to keep people who are not meant to be there, or to have access, out of the network and away from data. These processes cover the back doors and prevent the infiltration of malware. Authentication is the process of allowing permitted access through the front door of a network. Types of authentication include:
  - **Something you know**—passwords, PINs, challenge-response answers

    - **Something you have**—ID cards, smart cards
    - **Something you are**—biometrics (i.e. fingerprints, retinal scanners)
- **Authorization**—The authorization process used for access control requires that the system be able to identify and differentiate among users. Access rules (authorizations) specify who can access what. For example, access control is often based on least privilege, which means granting users only those accesses required to perform their duties. Access should be on a documented need-to-know and need-to-do (traditionally called Principle of Least Privilege) basis by type.

Computer access can be set for various levels (e.g., files, tables, data items, etc.). When information systems auditors review computer accessibility, they need to know what can be done with the access and what is restricted. For example, access restrictions at the file level generally include the following:

- Read only
- Write, create, update only
- Delete only
- Execute only
- A combination of the above

The least dangerous type of access is read-only, if the information being accessed is not sensitive or confidential. This is because the user cannot alter or use the computerized file beyond basic viewing or printing.

- **Accountability**—Accountability is the ability to map a given activity or event back to the responsible party. This aspect reinforces non-repudiation—that is, denying that the event or transaction occurred.
- **Encryption**—Encryption is the process of converting a plaintext message into a secure-coded form of text, called ciphertext. The ciphertext cannot be understood without converting back, via decryption—the reverse process—to plaintext. This is done via a mathematical function and a special encryption/decryption password called the key. In many countries, encryption is subject to governmental laws and regulations that limit the key size or define what may not be encrypted.

  Encryption is part of a broader science of secret languages called cryptography, which is generally used to:

    - Protect information stored on computers from unauthorized viewing and manipulation
    - Protect data in transit over networks from unauthorized interception and manipulation
    - Deter and detect accidental or intentional alterations of data
    - Verify authenticity of a transaction or document

  Encryption is limited since it cannot prevent the loss of data. It is possible to compromise encryption programs, if encryption keys are not protected adequately. Therefore, encryption should be regarded as an essential, but incomplete, form of access control that should be incorporated into an organization's overall computer security program.
- **Application Security**—Insecure applications open the organization to external attackers who may try to use unauthorized code to manipulate the application to access, steal, modify or delete sensitive data. Application security measures should be applied during the design and development phase of the application, followed by routine security countermeasures used throughout the life cycle.

2.5.5 Security Operations

Security operations includes several functions. Monitoring, detection and logging are integral parts of cybersecurity. With potential for attacks and data loss, it is necessary to monitor data and information flowing into and out of an organization. As this chapter illustrates, an organization can use a number of methods and tools to detect and log potential problems. Most of these methods revolve around the central concepts of ingress, egress and data loss prevention.

A log is a record of events that occur within the systems and networks of an organization. Logs are one of the most valuable tools to monitor controls and detect risk, but they are often underutilized. A log should contain a record of important events that occur on a system, and should include:

- Time of the event
- Changes to permissions
- System startup or shutdown
- Login or logout
- Changes to data
- Errors or violations
- Job failures

Log reviews can identify risk-relevant events such as compliance violations, suspicious behavior, errors, probes or scans and abnormal activity. A failure to review the logs can result in the organization not being aware of an ongoing attack. Logs may be required to comply with legislation and regulatory compliance and should also be preserved for forensic analysis, if needed, at a later time. IT professionals may find it useful to employ analysis tools when reviewing logs to filter pertinent data.

Ensuring proper segregation of duties (SoD) and time synchronization is particularly important when it comes to log files. The ability to change system configurations should be segregated from the ability to review, modify or delete logs in order to ensure that the organization can exercise proper oversight of administrative functions. Without time synchronization, it is extremely difficult to correlate information from different logs (server, router, firewall) to analyze an event's occurrence.

Common challenges relating to the effective use of logs include:

- Having too many data
- Difficulty in searching for relevant information
- Improper configuration (e.g., may not be enabled or contain appropriate data)
- Modification or deletion of data before they are read (e.g., too little storage space)

Page intentionally left blank

2.6 Chapter 2 Knowledge Check

REVIEW QUESTIONS

1. Number the following layers of the OSI model beginning with layer 1.

1.	Application
2.	Network
3.	Session
4.	Presentation
5.	Physical
6.	Transport
7.	Data link

2. How many bits are in a MAC address?

   A. 32
   B. 64
   C. 48
   D. 24

3. Match the following definitions to the correct information security objectives.

a. Guards against improper information modification or destruction.	1. Confidentiality
b. Preserves authorized restrictions on access or disclosure.	2. Availability
c. Ensures timely and reliable access to and use of information and systems.	3. Integrity

4. Which of the following network topologies connects computers using a drop cable (also called a T connector or a vampire clamp)?

   A. Ring
   B. Bus
   C. Star
   D. Mesh

5. What are the two portions of an IPv4 address?

Answers on page 104

Chapter 2 ANSWER KEY

Review Questions

1. Number the following layers of the OSI model beginning with layer 1. **Refer to page 72.**

1.	Physical
2.	Data link
3.	Network
4.	Transport
5.	Session
6.	Presentation
7.	Application

2. A. 32
   B. 64
   C. 48. Refer to page 80.
   D. 24

3. Match the following definitions to the correct information security objectives. **Refer to pages 94-96.**

a. Guards against improper information modification or destruction.	3. Integrity
b. Preserves authorized restrictions on access or disclosure.	1. Confidentiality
c. Ensures timely and reliable access to and use of information and systems.	2. Availability

4. A. Ring
   B. Bus. Refer to page 74.
   C. Star
   D. Mesh

5. Network ID, Host ID. **Refer to page 83.**

Glossary

A

Access control—The processes, rules and deployment mechanisms that control access to information systems, resources and physical access to premises

Access rights—The permission or privileges granted to users, programs or workstations to create, change, delete or view data and files within a system, as defined by rules established by data owners and the information security policy

Accountability—The ability to map a given activity or event back to the responsible party

Analog—A transmission signal that varies continuously in amplitude and time and is generated in wave formation

Scope Notes: Analog signals are used in telecommunications

Antivirus software—An application software deployed at multiple points in an IT architecture. It is designed to detect and potentially eliminate virus code before damage is done and repair or quarantine files that have already been infected.

Application layer—In the Open Systems Interconnection (OSI) communications model, the application layer provides services for an application program to ensure that effective communication with another application program in a network is possible.

Architecture—Description of the fundamental underlying design of the components of the business system, or of one element of the business system (e.g., technology), the relationships among them, and the manner in which they support enterprise objectives

Asset—Something of either tangible or intangible value that is worth protecting, including people, information, infrastructure, finances and reputation

Attack—An actual occurrence of an adverse event

Authentication—1. The act of verifying identity, i.e., user, system

Scope Notes: Can also refer to the verification of the correctness of a piece of data.

2. The act of verifying the identity of a user, the user's eligibility to access computerized information

Scope Notes: Authentication is designed to protect against fraudulent logon activity. It can also refer to the verification of the correctness of a piece of data.

Authenticity—Undisputed authorship

Availability—Ensuring timely and reliable access to and use of information

B

Backdoor—A means of regaining access to a compromised system by installing software or configuring existing software to enable remote access under attacker-defined conditions

Bandwidth—The range between the highest and lowest transmittable frequencies. It equates to the transmission capacity of an electronic line and is expressed in bytes per second or Hertz (cycles per second).

Biometrics—A security technique that verifies an individual's identity by analyzing a unique physical attribute, such as a handprint

Brute-force attack—Methodically trying all possible combinations of passwords or encryption keys until the correct one is found

Buffer overflow—Occurs when a program or process tries to store more data in a buffer (temporary data storage area) than it was intended to hold

Scope Notes: Because buffers contain a finite amount of data, the excess data can overflow into adjacent buffers, corrupting or overwriting the valid data held in them. Although it may occur accidentally through programming error, buffer overflow is an increasingly common type of security attack on data integrity. In buffer overflow attacks, the extra data may contain codes designed to trigger specific actions, in effect sending new instructions to the attacked computer that can damage user files, change data or disclose confidential information.

Business continuity plan (BCP)—A plan used by an enterprise to respond to disruption of critical business processes; depends on the contingency plan for restoration of critical systems

Byte—A sequence of adjacent bits, often an octet, operated on as a unit

C

Ciphertext—Information generated by an encryption algorithm to protect the plaintext and that is unintelligible to the unauthorized reader

Cloud computing—Convenient, scalable, on-demand network access to a shared pool of resources that can be rapidly provisioned and released with minimal management effort or service provider interaction

Compliance—Adherence to, and the ability to demonstrate adherence to, mandated requirements defined by laws and regulations, as well as voluntary requirements resulting from contractual obligations and internal policies

Confidentiality—Preserving authorized restrictions on access and disclosure, including means for protecting privacy and proprietary information

Control—The means of managing risk, including policies, procedures, guidelines, practices or organizational structures, which can be of an administrative, technical, management or legal nature

Scope Notes: Also used as a synonym for safeguard or countermeasure.

See also Internal control.

Countermeasure—Any process that directly reduces a threat or vulnerability

Criticality—The importance of a particular asset or function to the enterprise, and the impact if that asset or function is not available

Cryptography—The study of mathematical techniques related to aspects of information security, such as confidentiality, data integrity, entity authentication and data origin authentication

Cybersecurity—The protection of information assets by addressing threats to information processed, stored and transported by internetworked information systems

D

Data leakage—Unauthorized transmission of data from an organization either electronically or physically

Database—A collection of data, often with controlled redundancy, organized according to a schema to serve one or more applications. The data are stored so that they can be used by different programs without concern for the data structure or organization. A common approach is used to add new data and to modify and retrieve existing data. See Archival database.

Decryption—A technique used to recover the original plaintext from the ciphertext so that it is intelligible to the reader. The decryption is a reverse process of the encryption.

Denial-of-service attack (DoS)—An assault on a service from a single source that floods it with so many requests that it becomes overwhelmed and is either stopped completely or operates at a significantly reduced rate

Digital signature—An electronic identification of a person or entity using a public key algorithm that serves as a way for the recipient to verify the identity of the sender, integrity of the data and proof of transaction

Digital signature processor—Special processing unit specific to audio and telecommunication needs

Disaster—An emergency event of such great magnitude that it overwhelms the capacity to respond and takes considerable time from which to recover

Disaster recovery—Activities and programs designed to return the enterprise to an acceptable condition. The ability to respond to an interruption in services by implementing a disaster recovery plan (DRP) to restore an enterprise's critical business functions.

E

Egress—Network communications going out

Encapsulation—The technique used by layered protocols in which a lower-layer protocol accepts a message from a higher-layer protocol and places it in the data portion of a frame in the lower layer. In software development, it is a technique that isolates a system function or a set of data, and the operations on those data, within a module, and provides precise specifications for the module. See Abstraction, Information hiding and Software engineering.

Encryption—The process of taking an unencrypted message (plaintext), applying a mathematical function to it (encryption algorithm with a key) and producing an encrypted message (ciphertext).

Event—Something that happens at a specific place and/or time

Exploit—Method used to take advantage of a vulnerability

F

File Transfer Protocol (FTP)—A protocol used to transfer files over a Transmission Control Protocol/Internet Protocol (TCP/IP) network (Internet, UNIX, etc.).

Firewall—A system or combination of systems that enforces a boundary between two or more networks, typically forming a barrier between a secure and an open environment such as the Internet

G

Graphics processing unit—Special processing unit made to render high-quality images and video files

H

Hyperthreading—Intel propriety implementation of simultaneous multithreading

I

IEEE—Pronounced I-triple-E; IEEE is an organization composed of engineers, scientists and students. Best known for developing standards for the computer and electronics.

Impact—Magnitude of loss resulting from a threat exploiting a vulnerability

Incident—A violation or imminent threat of violation of computer security policies, acceptable use policies, guidelines or standard security practices

Information security—Ensures that, within the enterprise, information is protected against disclosure to unauthorized users (confidentiality), improper modification (integrity) and nonaccess when required (availability). Information security deals with all formats of information – paper documents, digital assets, intellectual property in people's minds, and verbal and visual communications.

Information systems (IS)—The combination of strategic, managerial and operational activities involved in gathering, processing, storing, distributing and using information and its related technologies

Scope Notes: Information systems are distinct from information technology (IT) in that an information system has an IT component that interacts with the process components.

Infrastructure as a Service (IaaS)—Offers the capability to provision processing, storage, networks and other fundamental computing resources, enabling the customer to deploy and run arbitrary software, which can include operating systems (OSs) and applications

Ingress—Network communications coming in

Intangible asset—An asset that is not physical in nature

Scope Notes: Examples include intellectual property (patents, trademarks, copyrights and processes), goodwill and brand recognition

Integrated circuit (IC)—Electronic circuit comprised of capacitors, transistors and resistors that is the building block of most electronic devices and equipment; also referred to as a chip or microchip

Integrity—The guarding against improper information modification or destruction, and includes ensuring information nonrepudiation and authenticity

Intellectual property—Intangible assets that belong to an enterprise for its exclusive use. Examples include patents, copyrights, trademarks, ideas and trade secrets.

Internal controls—The policies, procedures, practices and organizational structures designed to provide reasonable assurance that business objectives will be achieved and undesired events will be prevented or detected and corrected

Internet Control Message Protocol (ICMP)—A set of protocols that allow systems to communicate information about the state of services on other systems

Scope Notes: For example, ICMP is used in determining whether systems are up, maximum packet sizes on links, whether a destination host/network/port is available. Hackers typically use (abuse) ICMP to determine information about the remote site.

Internet Protocol (IP) packet spoofing— An attack using packets with the spoofed source Internet packet (IP) addresses

Scope Notes: This technique exploits applications that use authentication based on IP addresses. This technique also may enable an unauthorized user to gain root access on the target system.

Internet service provider (ISP)—A third party that provides individuals and enterprises with access to the Internet and a variety of other Internet-related services

Intruder—Individual or group gaining access to the network and its resources without permission

Investigation—The collection and analysis of evidence with the goal to identify the perpetrator of an attack or unauthorized use or access

IP address—A unique binary number used to identify devices on a TCP/IP network. May be IP version 4 or 6.

K

Kernel—Primary (of three) components of an operating system

L

Likelihood—The probability of something happening

Local area network (LAN)—Communication network that serves several users within a specified limited geographic area

Log—1. To record details of information or events in an organized record-keeping system, usually sequenced in the order in which they occurred

2. An electronic record of activity (e.g., authentication, authorization and accounting)

M

Malware—Short for malicious software. Designed to infiltrate, damage or obtain information from a computer system without the owner's consent. Examples of malware include computer viruses, worms, Trojan horses, spyware and adware.

Media access control (MAC)—Lower sublayer of the OSI Model Data Link layer

Media access control (MAC) address—A 48-bit unique identifier assigned to network interfaces for communications on the physical network segment

Microcontroller—Special processing unit useful in embedded systems, such as fleet vehicles and process control application

N

Need-to-Know—Principled approach of controlling what someone can see. Employees are only given access to data, systems or spaces that are necessary to do their job.

Network address—An identifier for a node or host on a telecommunications network

Nonrepudiation—The assurance that a party cannot later deny originating data; provision of proof of the integrity and origin of the data and that can be verified by a third party

Scope Notes: A digital signature can provide nonrepudiation.

O

Open Systems Interconnect (OSI) model—A seven-layer conceptual model that describes functions of computer network or telecommunication systems

Operating system—Software that controls the execution of programs and provides services, such as resource allocation, scheduling, input/output control and data management. Usually, operating systems are predominantly software, but partial or complete hardware implementations are possible.

P

Packet—Protocol data unit that is routed from source to destination in a packet-switched network

Scope Notes: A packet contains both routing information and data.

Password—A protected, generally computer-encrypted string of characters that authenticate a computer user to the computer system

Patch—Fixes to software programming errors and vulnerabilities

Personal identification number (PIN)—A type of password (i.e., a secret number assigned to an individual) that, in conjunction with some means of identifying the individual, serves to verify the authenticity of the individual

Scope Notes: PINs have been adopted by financial institutions as the primary means of verifying customers in an electronic funds transfer (EFT) system.

Phishing—A type of electronic mail (email) attack that attempts to convince a user that the originator is genuine, but with the intention of obtaining information for use in social engineering

Scope Notes: Phishing attacks may take the form of masquerading as a lottery organization advising the recipient or the user's bank of a large win; in either case, the intent is to obtain account and personal identification number (PIN) details. Alternative attacks may seek to obtain apparently innocuous business information, which may be used in another form of active attack.

PIN—See Personal identification number (PIN)

Plaintext—Digital information, such as cleartext, that is intelligible to the reader.

Platform as a Service (PaaS)—Offers the capability to deploy onto the cloud infrastructure customer-created or -acquired applications that are created using programming languages and tools supported by the provider

Port—A process or application-specific software element serving as a communication endpoint for the transport layer IP protocols (UDP and TCP)

Principle of least privilege (PoLP)—Principled approach of controlling what someone can do. Extension of need-to-know, whereby individuals are only granted the least amount of system access necessary to perform their jobs.

Printed circuit board (PCB)—The foundation of most electronic devices, onto which the electrical components, including semiconductors, connectors, resistors, capacitors, memory chips and processors, are mounted and linked via conductive copper circuits.

Privacy—The right of an individual to trust that others will appropriately and respectfully use, store, share and dispose of his/her associated personal and sensitive information within the context, and according to the purposes for which it was collected or derived

Probe—Inspect a network or system to find weak spots

Procedure—A document containing a detailed description of the steps necessary to perform specific operations in conformance with applicable standards. Procedures are defined as part of processes.

Protocol—The rules by which a network operates and controls the flow and priority of transmissions

R

Random access memory (RAM)—TA type of primary computer memory. RAM is volatile, and data is lost with power loss.

Read-only memory (ROM)—A type of primary computer memory. ROM is nonvolatile, and data stored there survives power loss.

Recovery—The phase in the incident response plan that ensures that affected systems or services are restored to a condition specified in the service delivery objectives (SDOs) or business continuity plan (BCP)

Regulation—Rules or laws defined and enforced by an authority to regulate conduct

Risk—The combination of the likelihood of an event and its impact

Risk tolerance—The acceptable level of variation that management is willing to allow for any particular risk as the enterprise pursues its objectives

Router—A networking device that can send (route) data packets from one local area network (LAN) or wide area network (WAN) to another, based on addressing at the network layer (Layer 3) in the open systems interconnection (OSI) model

Scope Notes: Networks connected by routers can use different or similar networking protocols. Routers usually are capable of filtering packets based on parameters, such as source addresses, destination addresses, protocol and network applications (ports).

S

Safeguard—A practice, procedure or mechanism that reduces risk

Secure Shell (SSH)—Network protocol that uses cryptography to secure communication, remote command line login and remote command execution between two networked computers

Segregation of duty (SoD)—See Segregation/separation of duties (SoD).

Semiconductor—Substrate for integrated circuit that regulates electric current and is often made primarily of silicon

Sensitivity—A measure of the impact that improper disclosure of information may have on an enterprise

Social engineering—An attack based on deceiving users or administrators at the target site into revealing confidential or sensitive information

Spear phishing—An attack designed to entice specific individuals or groups to obtain important information, where social engineering techniques are used to masquerade as a trusted party to obtain important information, such as passwords from the victim

Spoofing—Faking the sending address of a transmission in order to gain illegal entry into a secure system

T

Target—Person or asset selected as the aim of an attack

Threat—Anything (e.g., object, substance, human) that is capable of acting against an asset in a manner that can result in harm

Scope Notes: A potential cause of an unwanted incident (ISO/IEC 13335)

Threat agent—Methods and things used to exploit a vulnerability

Scope Notes: Examples include determination, capability, motive and resources.

Threat event—Any event during which a threat element/actor acts against an asset in a manner that has the potential to directly result in harm

Topology—The physical layout of how computers are linked together

Scope Notes: Examples of topology include ring, star and bus.

Transmission Control Protocol (TCP)—A connection-based Internet protocol that supports reliable data transfer connections

Scope Notes: Packet data are verified using checksums and retransmitted if they are missing or corrupted. The application plays no part in validating the transfer.

Transmission Control Protocol Internet Protocol (TCP/IP)—Provides the basis for the Internet; a set of communication protocols that encompass media access, packet transport, session communication, file transfer, electronic mail (email), terminal emulation, remote file access and network management

U

User Datagram Protocol (UDP)—A connectionless Internet protocol that is designed for network efficiency and speed at the expense of reliability

V

Virtualization—The process of adding a guest application and data onto a virtual server, recognizing that the guest application will ultimately be removed from this physical server

Voice-over Internet Protocol (VoIP)—Also called IP Telephony, Internet Telephony and Broadband Phone, a technology that makes it possible to have a voice conversation over the Internet or over any dedicated Internet Protocol (IP) network instead of over dedicated voice transmission lines

Volatile data—Data that change frequently and can be lost when the system power is shut down

Vulnerability—A weakness in the design, implementation, operation or internal control of a process that could expose the system to adverse threats from threat events

W

Wide area network (WAN)—A computer network connecting multiple offices or buildings over a larger area

Z

Zero-day exploit—A vulnerability that is exploited before the software creator/vendor is even aware of its existence. May also refer to known flaws that do not have a patch available.

9781604206562